Progress in IS

More information about this series at http://www.springer.com/series/10440

Claus-Peter H. Ernst
Editor

The Drivers of Wearable Device Usage

Practice and Perspectives

 Springer

Editor
Claus-Peter H. Ernst
Frankfurt University of Applied Sciences
Frankfurt am Main
Germany

ISSN 2196-8705   ISSN 2196-8713   (electronic)
Progress in IS
ISBN 978-3-319-30374-1   ISBN 978-3-319-30376-5   (eBook)
DOI 10.1007/978-3-319-30376-5

Library of Congress Control Number: 2016932345

Printed on acid-free paper

This Springer imprint is published by SpringerNature
The registered company is Springer International Publishing AG Switzerland

Preface

One might argue that technology usage and adoption are among the most over-studied topics in the IS literature. Indeed, I agree that there is no need for yet another paper that focuses solely on the influence of usefulness and enjoyment on technologies. However, I believe that the usage and adoption of *specific* technologies might be driven by specific factors.

In line with this, this book collects six studies that research specific kinds of wearable devices, such as smartwatches and smartglasses, and analyzes the factors that could be driving their usage. As a result, this book provides researchers with important insights into the specifics of technology usage and also holds specific implications for practitioners such as manufacturers of wearable devices.

I would like to thank the authors of each chapter: Daniel Weiz, Gagat Anand, Bastian Stock, Tiago Patrick dos Santos Ferreira, Florian Rheingans, Burhan Cikit, Frederik Mayer, Duc Nguyen, Alexander Ernst, Patrick Reinelt, and Shewit Hadish.

Finally, my special thanks go to my friend and mentor Franz Rothlauf, who helps, supports, and continues to teach me so much: without him, I never would have done any scientific work at all.

Frankfurt am Main   Claus-Peter H. Ernst
January 2016

Contents

The Influence of Subjective Norm on the Usage of Smartglasses

Daniel Weiz, Gagat Anand and Claus-Peter H. Ernst

Abstract One factor hindering people's usage of smartglasses seems to be that of Subjective Norm. More specifically, there are multiple reports of people using Google Glass being criticized in public, due to the general public's perception that their privacy is at risk because of the device's integrated recording functionalities. In this article, we empirically evaluate the influence of Subjective Norm on smartglasses usage. After collecting 111 completed online questionnaires about one specific pair of smartglasses, Google Glass, and applying a structural equation modeling approach, our findings indicate that smartglasses are at least partly utilitarian technologies whose usage is influenced by Perceived Usefulness. Furthermore, although we could not confirm a direct positive influence of Subjective Norm on the Actual System Use of smartglasses, we confirmed an indirect positive influence of Subjective Norm on Actual System Use through Perceived Usefulness. These findings suggest that smartglasses manufacturers need to emphasize the instrumental benefits of their devices. In addition, the manufacturers need to address users' potential negative perceptions of smartglasses stemming from users' beliefs that the general public has a negative opinion of the device.

1 Introduction

After notebooks, smartphones and tablets, wearable devices—i.e., "electronic technologies or computers that are incorporated into items of clothing and accessories which can comfortably be worn on the body" (Tehrani and Andrew 2014)—might be the next big thing in mobile computing. There is a broad range of different kinds of wearable devices, from bracelets that measure people' daily activities to smartglasses that enhance the real world with virtual functions or immerse the user

D. Weiz · G. Anand · C.-P.H. Ernst (✉)
Frankfurt University of Applied Sciences, Frankfurt am Main, Germany
e-mail: cernst@fb3.fra-uas.de

© Springer International Publishing Switzerland 2016
C.-P.H. Ernst (ed.), *The Drivers of Wearable Device Usage*,
Progress in IS, DOI 10.1007/978-3-319-30376-5_1

into fully virtual worlds. According to forecasts, the revenues of wearable devices are expected to exceed 9 billion Euros in 2018 in Europe alone (Statista 2015). However, some companies seem to have problems with successfully bringing their wearable device products to market. Smartglasses in particular seem to suffer from a lack of acceptance by a great majority of people. Indeed, Google ended their beta program for their Google Glass in the beginning of 2015 (Lardinois 2015).

One factor hindering people's acceptance of smartglasses might be Subjective Norm. More specifically, there are multiple reports of people using Google Glass being criticized in public, due to the general public's perception that their privacy was threatened by the device's integrated camera and microphone (cf. Villapaz 2015). Since Subjective Norm has proven to be particularly important for the acceptance and usage behavior of new kinds of technology such as wearable devices (Watjatrakul 2013), this lack of general social acceptance might constitute a serious problem for smartglasses manufacturers.

In this article, we seek to shed light on this potential influence factor of smartglasses acceptance by carrying out an empirical evaluation. After collecting 111 completed online questionnaires about one specific pair of smartglasses, Google Glass, and applying a structural equation modeling approach, our findings indicate that smartglasses are at least partly utilitarian technologies whose usage is influenced by Perceived Usefulness. Furthermore, although we could not confirm a direct positive influence of Subjective Norm on Actual System Use, we confirmed an indirect positive influence of Subjective Norm on Actual System Use through Perceived Usefulness. These findings suggest that smartglasses manufacturers need to emphasize the instrumental benefits of their devices. In addition, the manufacturers need to address users' potential negative perceptions of smartglasses stemming from users' beliefs that the general public has a negative opinion of the device.

In the next section, we will present background information on smartglasses, introduce Perceived Usefulness as an influence factor of utilitarian technologies, and also present the theoretical foundations of Subjective Norm. Following this, we will present our research model and research design. We will then reveal and discuss our results before summarizing our findings, presenting their theoretical as well as practical implications, and providing an outlook on further research.

2 Theoretical Background

2.1 Smartglasses

Smartglasses are head-mounted displays and can be divided into two categories: Augmented Reality Smartglasses (ARSG) and Virtual Reality Smartglasses (VRSG) (Amorim et al. 2013; Due 2014; Milgram et al. 1994; Nilsson and

Johansson 2007). VRSG such as Oculus Rift place the user in an artificial environment. The user cannot interact with the real world—rather, he/she is completely immersed in a virtual world. In contrast, ARSG such as Google Glass allow users to interact with the real world since they complement the real world with virtual functions (Azuma 1997; Nilsson and Johansson 2007).

Multiple instrumental benefits of smartglasses have been discussed in the literature. For example, one study evaluated the use of these devices by people with Parkinson's disease and found that they might be useful in helping them carry out everyday tasks (McNaney et al. 2014). In addition, smartglasses might prove useful for insurance companies by providing a direct connection to a crash scene and the ensuing damage, which insurance specialists can then analyze (Kim et al. 2013). Overall, smartglasses are expected to improve productivity, offer new ways to visualize problems and solutions, and enhance collaboration (Nguyen 2013), making them at least partly utilitarian technologies.

2.2 The Role of Perceived Usefulness on Smartglasses Usage

Generally, utilitarian technologies "aim to provide instrumental value to the user" (Van der Heijden 2004, p. 696). Perceived Usefulness—"the degree to which a person believes that using a particular system would enhance his or her job [and task] performance" (Davis 1989, p. 320)—centers on the motivations and benefits that are external to the system-user interaction itself, referred to as extrinsic motivations (Brief and Aldag 1977; Van der Heijden 2004). For example, the external benefits/extrinsic motivations of a text-processing program can be to foster a good writing performance in terms of a well-structured and orthographically error-free text (Davis et al. 1989).

Various studies in a variety of contexts have consistently confirmed that Perceived Usefulness is a central antecedent of utilitarian technology usage (e.g., Davis 1989). By applying these findings to our context, a person can be expected to use smartglasses if he/she believes that they fulfill his expectations with regards to instrumental benefits, that is to their Perceived Usefulness.

2.3 Subjective Norm

The Theory of Reasoned Action (Fishbein and Ajzen 1975) postulates that beliefs regarding the outcome of a specific behavior (such as Perceived Usefulness) as well as Subjective Norm indirectly influence the Actual Behavior of people.

Subjective Norm denotes social influences and can be defined as the degree to which a persons believes "that most people who are important to him think he should or should not perform the behavior in question" (Fishbein and Ajzen 1975,

p. 302). It has proven to be particularly important for the acceptance and usage behavior of new kinds of technology such as wearable devices (Watjatrakul 2013).

Moreover, some evidence suggests that the perception of lack of social acceptance might be a problem for smartglasses usage in particular. In fact, multiple news stories report that people using Google Glass have been publicly critized, or even attacked (cf. Villapaz 2015). The lack of general social acceptance of these devices might thus constitute a problem for smartglasses manufacturers.

3 Research Model

In the following section, we will present our research model in Fig. 1 and then outline our corresponding hypotheses.

As described earlier, smartglasses are useful in a variety of fields: for example, they can enhance collaboration and offer new ways of visualizing problems and solutions (Nguyen 2013). Therefore, smartglasses are at least partly utilitarian technologies (cf. Ernst et al. 2013) that provide users with benefits that are external to the system-user interaction itself. Perceived Usefulness is commonly accepted to be an important antecedent of utilitarian technologies' Actual System Use (e.g., Davis et al. 1989). We hypothesize that:

H1   There is a positive influence of Perceived Usefulness on the Actual System Use of smartglasses.

Fishbein and Ajzen (1975) postulate that Subjective Norm is an important influence factor of people's behavior. In fact, a great number of theories point out the importance of social aspects in terms of people's acceptance of technology. These include critical mass (Markus 1990), social influence (Fulk et al. 1987), adaptive structuration (Poole and DeSanctis 1990), hermeneutic interpretation (Lee 1994), and critical social theory (Ngwenyama and Lee 1997). However, empirical studies have found mixed evidence regarding the role of Subjective Norm on technology usage (Venkatesh and Morris 2000): Some studies did not even include Subjective Norm (e.g., Adams et al. 1992; Szajna 1994; Szajna 1996); Some found that Subjective Norm's influence on Actual System Use was significant (e.g., Hartwick and Barki

Fig. 1 Research model

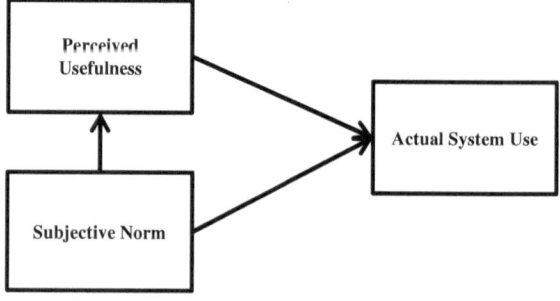

1994), and some found it to be non-significant (e.g., Davis et al. 1989). However, in the context of smartglasses, there might be a strong influence of Subjective Norm on Actual System Use since people cannot hide their usage in public—indeed, they need to wear the device overtly on their face. We hypothesize that:

H2 *There is a positive influence of Subjective Norm on the Actual System Use of smartglasses.*

Moreover, multiple studies have emphasized that not only people's behavior, but also their beliefs, can be influenced by others (e.g., Burnkrant and Cousineau 1975; Deutsch and Gerard 1955). In line with this, multiple studies (e.g., Schepers and Wetzels 2007; Shen et al. 2006) have found a significant positive influence of Subjective Norm on people's beliefs regarding a system's instrumental benefits, i.e., its Perceived Usefulness. We hypothesize that:

H3 *There is a positive influence of Subjective Norm on the Perceived Usefulness of smartglasses.*

4 Research Design

4.1 Data Collection

To empirically evaluate our research model, we collected 111 completed questionnaires by publishing an English-language online survey about one specific pair of smartglasses, Google Glass, in multiple corresponding communities on Google+ as well as on reddit. At the beginning of the questionnaire, we provided a short description of Google Glass, including official images and an explanation of its general functionalities. Google Glass, which was only sold for a limited time as a public beta product to so-called explorers in the UK and US (as of June 2015), promised to add "an augmented-reality overlay to whatever ... [users are] looking at, automatically bringing up relevant information from various Google sources" (Engadget 2015).

99 of our respondents were male (89.19 %) and 12 were female (10.81 %). The average age was 33.23 years (standard deviation: 8.78). 3 respondents were unemployed (2.70 %), 86 were currently employed (77.48 %), 17 were students (15.32 %), and 5 selected "other" as a description of themselves (4.50 %).

4.2 Measurement

We adapted existing reflective scales to our context in order to measure Actual System Use, Perceived Usefulness, and Subjective Norm (Alarcón-del-Amo et al.

Table 1 Items of our measurement model

Construct	Item	Source
Actual System Use	On average, how often do you use Google Glass? (AU1)	Davis et al. (1989)
	How frequently do you use Google Glass? (AU2)	
Perceived Usefulness	Google Glass benefits me (PU1)	Alarcón-del-Amo et al. (2012) cf. Ernst et al. (2013)
	Google Glass is an effective tool (PU2)	
	I consider that Google Glass is useful to me (PU3)	
Subjective Norm	People who influence my behavior think I should use Google Glass (SN1)	Taylor and Todd (1995)
	Others think I should use Google Glass (SN2)	
	People who are important to me think I should use Google Glass (SN3)	

2012; Davis et al. 1989; Taylor and Todd 1995). Table 1 presents the resulting reflective items with their corresponding sources. Actual System Use was measured in the same manner as in Davis et al. (1989, p. 991), and all other items were measured using a seven-point Likert-type scale ranging from "strongly agree" to "strongly disagree".

5   Results

Since our data was not distributed joint multivariate normal (cf. Hair et al. 2011), we used the Partial-Least-Squares approach via SmartPLS 3.2.0 (Ringle et al. 2015). With 111 datasets, we met the suggested minimum sample size threshold of "ten times the largest number of structural paths directed at a particular latent construct in the structural model" (Hair et al. 2011, p. 144). To test for significance, we used the integrated Bootstrap routine with 5,000 samples (Hair et al. 2011).

In the following section, we will evaluate our measurement model. Indeed, we will examine the indicator reliability, the construct reliability, and the discriminant validity of our reflective constructs. Finally, we will present the results of our structural model.

5.1   Measurement Model

Tables 2 and 3 present the correlations between constructs along with the Average Variance Extracted (AVE) and Composite Reliability (CR), and our reflective items' factor loadings, respectively: All items loaded high (0.813 or more) and significant ($p < 0.001$) on their parent factor and, hence, met the suggested threshold

Table 2 Correlations between constructs [AVE (CR) on the diagonal]

	AU	PU	SN
Actual System Use (AU)	0.997 (0.998)		
Perceived Usefulness (PU)	0.305	0.740 (0.895)	
Subjective Norm (SN)	0.275	0.509	0.919 (0.971)

Table 3 Reflective items' loadings (T-Values)

	AU	PU	SN
AU1	0.999 (821.7)	0.309	0.277
AU2	0.998 (768.5)	0.299	0.271
PU1	0.320	0.888 (40.4)	0.549
PU2	0.189	0.813 (14.7)	0.280
PU3	0.243	0.877 (39.2)	0.412
SN1	0.238	0.431	0.951 (76.2)
SN2	0.252	0.512	0.956 (100.1)
SN3	0.296	0.513	0.969 (145.5)

of indicator reliability of 0.70 (Hair et al. 2011); AVE and CR were higher than 0.74 and 0.89, respectively, meeting the suggested construct reliability thresholds of 0.50/0.70 (Hair et al. 2009). The loadings from our reflective indicators were highest for each parent factor and the square root of the AVE of each construct was larger than the absolute value of the construct's correlations with its counterparts, thus indicating discriminant validity (Fornell and Larcker 1981; Hair et al. 2011).

5.2 Structural Model

Figure 2 presents the path coefficients of the previously hypothesized relationships as well as the R^2s of both endogenous variables (* = $p < 0.05$; *** = $p < 0.001$; ns = non-significant). Hypothesis 2 was not confirmed since Subjective Norm, in line with the findings of other similar studies (e.g., Davis et al. 1989), had no

Fig. 2 Findings

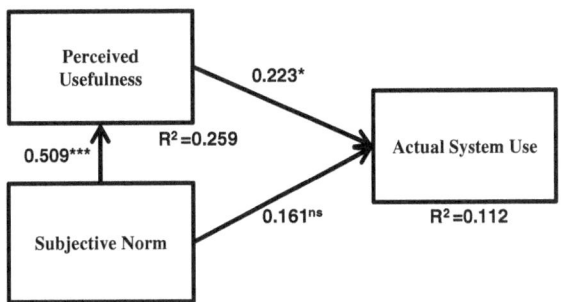

significant influence on Actual System Use ($\beta = 0.161$, $t = 1.391$). However, Subjective Norm was found to have a significant influence on Perceived Usefulness ($\beta = 0.509$, $p < 0.001$), which, in turn, was found to have a positive influence on Actual System Use ($\beta = 0.223$, $p < 0.05$), confirming hypotheses 3 and 1, respectively.

Our research model included two predecessors of Actual System Use (Perceived Usefulness and Subjective Norm), and one predecessor of Perceived Usefulness (Subjective Norm). By taking this into account, the explanatory power of our structural model is good, since it explains 11.2 % of the variances of Actual System Use as well as 25.9 % of the variances of Perceived Usefulness.

In summary, our findings indicate that smartglasses are at least partly utilitarian technologies whose usage is influenced by Perceived Usefulness. Furthermore, although we could not confirm a direct positive influence of Subjective Norm on Actual System Use, our findings indicate an indirect positive influence of Subjective Norm on Actual System Use through Perceived Usefulness.

6 Conclusion

In this article, we evaluated the potential influence of Subjective Norm on smartglasses usage. After collecting 111 completed online questionnaires about one specific pair of smartglasses, Google Glass, and applying a structural equation modeling approach, our findings indicate that smartglasses are at least partly utilitarian technologies whose usage is influenced by Perceived Usefulness. Furthermore, although we could not confirm a direct positive influence of Subjective Norm on the Actual System Use of smartglasses, we confirmed an indirect positive influence of Subjective Norm on Actual System Use through Perceived Usefulness.

Our findings have important practical implications. Indeed, they suggest that smartglasses manufacturers need to emphasize the instrumental benefits of their devices. In addition, the manufacturers need to address users' potential negative perceptions of smartglasses stemming from users' beliefs that the general public has a negative opinion of the device.

For example, the manufacturers could choose to design them to be less intrusive by removing the camera functionality, or by adapting the physical aspect of the device so that it appears to be less intrusive. In fact, Google has already worked with the Ray-Ban brand in order to create glasses attachments that enable people to attach real sunglasses to the Google Glass frame (Lawler 2014). This could help conceal the technology from others.

Our study has some limitations. First, our empirical findings are based on only one specific pair of smartglasses: Google Glass. Hence, the results found for these particular smartglasses might be very different in studies that use other smartglasses. Moreover, our sample contained mostly men and we also only surveyed English-speaking people. Thus, our results might be potentially less representative for women and might also not hold true for non-English speakers. Furthermore, our findings might be biased since we posted our survey exclusively to communities that were interested in Google Glass. Also, our sample individuals were relatively young (mean: 33.23 years; standard deviation: 8.78). Hence, differences might be found for other age groups. Finally, our survey was only conducted online and, hence, excluded people that do not use the Internet (which could also explain the lack of older people in our sample).

As a next step, we would like to equip people in Germany with a Google Glass device for one week with specific tasks to perform, such as the task of navigating to different points of interest. One group of participants would use a normal Google Glass, another group would use a Google Glass with a Ray-Ban add-on in order to make it less obvious that it is a technological device. Before and after this hands-on phase, we plan to let our study participants fill out a German-language question-naire in order to analyze for differences between the groups with regards to Subjective Norm and Perceived Usefulness, as well as for potential changes in perception through real-world usage.

References

Adams, D. A., Nelson, R. R., & Todd, P. A. (1992). Perceived usefulness, ease of use and usage of information technology: A replication. *MIS Quarterly, 16*(2), 227–250.

Alarcón-del-Amo, M.-C., Lorenzo-Romero, C., & Gomez-Borja, M.-A. (2012). Analysis of acceptance of social networking sites. *African Journal of Business Management, 6*(29), 8609–8619.

Amorim, J. A., Matos, C., Cuperschmid, A. R. M., Gustavsson, P. M., & Pozzer, C. T. (2013). Augmented reality and mixed reality technologies: Enhancing training and mission preparation with simulations. In *STO Modelling and Simulation Group Conference 2013 Proceedings.* Paper 111.

Azuma, R. T. (1997). A survey of augmented reality. *Presence: Teleoperators and Virtual Environments, 6*(4), 355–385.

Brief, A. P., & Aldag, R. J. (1977). The intrinsic-extrinsic dichotomy: Toward conceptual clarity. *Academy of Management Review, 2*(3), 496–500.

Burnkrant, R. E., & Cousineau, A. (1975). Informational and normative social influence in buyer behavior. *Journal of Consumer Research, 2*(3), 206–215.

Davis, F. D. (1989). Perceived usefulness, perceived ease of use, and user acceptance of information technology. *MIS Quarterly, 13*(2), 319–340.

Davis, F. D., Bagozzi, R. P., & Warshaw, P. R. (1989). User accaptance of computer technology: A comparsion of two theoretical models. *Management Science, 35*(8), 982–1003.

Deutsch, M., & Gerard, H. B. (1955). A study of normative and informational social influences upon individual judgment. *The Journal of Abnormal and Social Psychology, 51*(3), 629–636.

Due, B. L. (2014). The future of smart glasses: An essay about challenges and possibilities with smart glasses. *Working Papers on Interaction and Communication 1*(2), 1–21 (University of Copenhagen).

Engadget. (2015). Google Glass review. http://www.engadget.com/products/google/glass. Accessed July 22, 2015.

Ernst, C.-P. H., Pfeiffer, J. & Rothlauf, F. (2013). Hedonic and utilitarian motivations of social network site adoption. Johannes Gutenberg University Mainz, Working Paper.

Fishbein, M., & Ajzen, I. (1975). *Belief, attitude, intention and behavior: An introduction to theory and research*. Reading, Mass: Addison-Wesley.

Fornell, C., & Larcker, D. F. (1981). Evaluating structural equation models with unobservable variables and measurement error. *Journal of Marketing Research, 18*(1), 39–50.

Fulk, J., Steinfield, C. W., Schmitz, J., & Power, J. G. (1987). A social information processing model of media use in organizations. *Communications Research, 14*(5), 529–552.

Hair, J. F., Black, W. C., Babin, B. J., & Anderson, R. E. (2009). *Multivariate data analysis* (7th ed.). Upper Saddle River: Prentice Hall.

Hair, J. F., Ringle, C. M., & Sarstedt, M. (2011). PLS-SEM: Indeed a silver bullet. *Journal of Marketing Theory and Practice, 19*(2), 139–151.

Hartwick, J., & Barki, H. (1994). Explaining the role of user participation in information system use. *Management Science, 40*(4), 440–465.

Kim, M., Francis, A., Gupta, R., & Kumar, M. (2013): Google Glass: Insurance's next killer app. Cognizant White Paper.

Lardinois, F. (2015). Google Glass explorer program shuts down, team now reports to Tony Fadell. http://techcrunch.com/2015/01/15/google-glass-exits-x-labs-as-explorer-program-shuts-down-team-now-reports-to-tony-fadell. Accessed June 29, 2015.

Lawler, R. (2014). Ray-Ban and Oakley are working with Google Glass. http://www.engadget.com/2014/03/24/google-glass-ray-ban-oakley-luxottica. Accessed June 29, 2015.

Lee, A. (1994). Electronic mail as a medium for rich communication: An empirical investigation using hermeneutic interpretation. *MIS Quarterly, 18*(2), 143–157.

Markus, M. L. (1990). Toward a 'critical mass' theory of interactive media. In J. Fulk & C. Steinfield (Eds.), *Organizations and Communication Technology* (pp. 491–511). Newbury Park, CA: Sage.

McNaney, R., Vines, J., Roggen, D., Balaam, M., Zhang, P., Poliakov, I., et al. (2014). Exploring the acceptability of Google Glass as an everyday assistive device for people with parkinson's. In *CHI 2014 Proceedings* (pp. 2551–2554).

Milgram, P., Takemura, H., Utsumi, A., & Kishino, F. (1994). Augmented reality: A class of displays on the reality-virtuality continuum. In *Telemanipulator and Telepresence Technologies 1994 Proceedings* (pp. 282–292).

Nguyen, T. H. (2013). Innovation insight: Augmented reality will become an important workplace tool. https://www.gartner.com/doc/2640230/innovation-insight-augmented-reality-important. Accessed June 16, 2015.

Ngwenyama, O. J., & Lee, A. S. (1997). Communication richness in electronic mail: Critical social theory and the contextuality of meaning. *MIS Quarterly, 21*(2), 145–168.

Nilsson, S., & Johansson, B. (2007). Fun and usable: Augmented reality instructions in a hospital setting. In *Australasian Conference on Computer-Human Interaction 2007 Proceedings* (pp. 123–130).

Poole, M. S., & DeSanctis, G. (1990). Understanding the use of group decision support systems: The theory of adaptive structuration. In J. Fulk & C. Steinfield (Eds.), *Organizations and Communication Technology* (pp. 173–191). Newbury Park, CA: Sage.

Ringle, C. M., Wende, S., & Becker, J.-M. (2015). SmartPLS 3. http://www.smartpls.com.

Schepers, J., & Wetzels, M. (2007). A meta-analysis of the technology acceptance model: Investigating subjective norm and moderation effects. *Information and Management, 44*(1), 90–103.

Shen, D., Laffey, J., Lin, Y., & Huang, X. (2006). Social influence for perceived usefulness and ease-of-use of course delivery systems. *Journal of Interactive Online Learning, 5*(3), 270–282.

Statista. (2015). Prognose zum Umsatz mit Wearable Technology in Europa von 2013 bis 2018 (in Milliarden Euro). http://de.statista.com/statistik/daten/studie/322222/umfrage/prognose-zum-umsatz-mit-wearable-computing-geraeten-in-europa. Accessed June 29, 2015.

Szajna, B. (1994). Software evaluation and choice: Predictive validation of the technology acceptance instrument. *MIS Quarterly, 18*(3), 319–324.

Szajna, B. (1996). Empirical evaluation of the revised technology acceptance model. *Management Science, 42*(1), 85–92.

Taylor, S., & Todd, P. A. (1995). Understanding information technology usage: A test of competing models. *Information Systems Research, 6*(2), 144–176.

Tehrani, K., & Andrew, M. (2014). Wearable technology and wearable devices: Everything you need to know. http://www.wearabledevices.com/what-is-a-wearable-device. Accessed June 29, 2015.

Van Der Heijden, H. (2004). User acceptance of hedonic information systems. *MIS Quarterly, 28*(4), 695–704.

Venkatesh, V., & Morris, M. G. (2000). Why don't men ever stop to ask for directions? Gender, social influence, and their role in technology acceptance and usage behavior. *MIS Quarterly, 24*(1), 115–139.

Villapaz, L. (2015). What Google needs to do to fix glass and end the 'Glasshole' stigma. http://www.ibtimes.com/what-google-needs-do-fix-glass-end-glasshole-stigma-1790398. Accessed June 29, 2015.

Watjatrakul, B. (2013). Intention to use a free voluntary service. The effects of social influence, knowledge and perceptions. *Journal of Systems and Information Technology, 15*(2), 202–220.

Does Perceived Health Risk Influence Smartglasses Usage?

Bastian Stock, Tiago Patrick dos Santos Ferreira
and Claus-Peter H. Ernst

Abstract The World Health Organization has warned populations about illnesses that can develop due to radiation. Since smartglasses, which are worn on the head right next to the brain, can emit radiation, their usage might be hindered by the Perceived Health Risks people associate with such devices. In this article, we empirically evaluate the topic by studying the influence of Perceived Health Risk on smartglasses usage. After collecting 109 completed online questionnaires about one specific pair of smartglasses, Microsoft HoloLens, and applying a structural equation modeling approach, our findings indicate that smartglasses are at least partly hedonic technologies whose usage is influenced by Perceived Enjoyment. Furthermore, although we could not confirm a direct negative influence of Perceived Health Risk on the Behavioral Intention to Use smartglasses, we confirmed an indirect negative influence of Perceived Health Risk on Behavioral Intention to Use through Perceived Enjoyment. These findings suggest that smartglasses manufacturers need to emphasize the hedonic benefits of their devices as well as address people's potential negative perceptions of these devices in terms of their health.

1 Introduction

After notebooks, smartphones and tablets, wearable devices—i.e., "electronic technologies or computers that are incorporated into items of clothing and accessories which can comfortably be worn on the body" (Tehrani and Andrew 2014)—might be the next driver of mobile computing. There is a broad range of different kinds of wearable devices, from bracelets that measure people' daily activities to smartglasses that enhance the real world with virtual functions or immerse the user into fully virtual worlds. According to forecasts, the revenues of wearable devices are expected to exceed 9 billion Euros in 2018 in Europe alone (Statista 2014).

B. Stock · T.P. dos Santos Ferreira · C.-P.H. Ernst (✉)
Frankfurt University of Applied Sciences, Frankfurt am Main, Germany
e-mail: cernst@fb3.fra-uas.de

© Springer International Publishing Switzerland 2016
C.-P.H. Ernst (ed.), *The Drivers of Wearable Device Usage*,
Progress in IS, DOI 10.1007/978-3-319-30376-5_2

However, some companies seem to have problems with successfully bringing their wearable devices to market. Smartglasses in particular, which are overtly worn on the head, seem to suffer from a lack of acceptance by a majority of people. Indeed, Google ended their explorer program for their Google Glass in the beginning of 2015 (Lardinois 2015).

One aspect hindering people's acceptance of smartglasses might be the Perceived Health Risk associated with these devices. More specifically, smartglasses can emit radiation, which might have negative health consequences such as the development of illnesses (Burgess 2002; Seigneur et al. 2010). The potential fear of people regarding their health might be increasingly enhanced by to the fact that smartglasses are worn right next to the brain, thus exposing their brains to potentially harmful radiation.

In this article, we seek to shed light on the potential role of Perceived Health Risk on the usage of smartglasses by carrying out an empirical evaluation. After collecting 109 completed online questionnaires about one specific pair of smartglasses, Microsoft HoloLens, and applying a structural equation modeling approach, our findings indicate that smartglasses are at least partly hedonic technologies whose usage is influenced by Perceived Enjoyment. Furthermore, although we could not confirm a direct negative influence of Perceived Health Risk on the Behavioral Intention to Use smartglasses, we confirmed an indirect negative influence of Perceived Health Risk on Behavioral Intention to Use through Perceived Enjoyment. These findings suggests that smartglasses manufacturers need to emphasize the hedonic benefits of their devices as well as to address people's potential negative perceptions of smartglasses in terms of their health.

In the next section, we will present background information on smartglasses, introduce Perceived Enjoyment as an influence factor of hedonic technologies, and also present the theoretical foundations of the Perceived Health Risk construct. Following this, we will present our research model and research design. We will then reveal and discuss our results before summarizing our findings, presenting their theoretical as well as practical implications, and providing an outlook on further research.

2 Theoretical Background

2.1 Smartglasses

Smartglasses are head-mounted displays and can be divided into two categories: Augmented Reality Smartglasses (ARSG) and Virtual Reality Smartglasses (VRSG) (Amorim et al. 2013; Due 2014; Milgram et al. 1994; Nilsson and Johansson 2007). VRSG such as Oculus Rift place the user in an artificial environment. The user cannot interact with the real world—rather, he/she is completely immersed in the virtual world. In contrast, ARSG such as Google Glass allow users to interact with the real world since they complement the real world with virtual functions (Azuma 1997; Nilsson and Johansson 2007).

Multiple instrumental benefits of smartglasses have been discussed in the literature (e.g., Kim et al. 2013; McNaney et al. 2014; Nguyen 2013). Additionally, smartglasses can also be used for hedonic purposes such as video games and some are even built exclusively for gaming purposes. However, due to the novelty of these devices, only a few studies have studied the factors that drive people to use smartglasses (e.g., Rauschnabel et al. 2015).

2.2 The Role of Perceived Enjoyment on Smartglasses Usage

Generally, hedonic technologies "aim to provide self-fulfilling value to the user, ... [which] is a function of the degree to which the user experiences fun when using the system" (Van der Heijden 2004, p. 696). Perceived Enjoyment—"the extent to which the activity of using a specific system is perceived to be enjoyable in its own right, aside from any performance consequences resulting from system use" (Venkatesh 2000, p. 351)—reflects a hedonic system's intrinsic motivations, such as fun, enjoyment, and other positive experiences, which stem directly from the system-user interaction (Brief and Aldag 1977; Van der Heijden 2004; Venkatesh et al. 2012).

Various studies in a variety of contexts have consistently confirmed that Perceived Enjoyment is a central antecedent of hedonic technology usage (e.g., Van der Heijden 2004). By applying these findings to our contexts, a person can be expected to use smartglasses if he/she believes that they fulfill his expectations with regards to enjoyment.

2.3 Perceived Health Risk

Risk can be generally described as "the extent to which there is an uncertainty in significant and disappointing outcomes that may be realized" (Chen 2013, p. 1222; Sitkin and Pablo 1992). Perceived Risk is thus consistently understood as "the expectation of losses associated with... [specific actions]" (Peter and Ryan 1976, p. 185).

Several studies have confirmed that Perceived Risk (e.g., Privacy Risk) can exert a negative influence on the usage of technologies (e.g., Egea and Gonzáles 2010; Tan 1999). Moreover, in addition to a direct effect of Perceived Risk on technology usage, Ernst (2014) also confirmed that Perceived Risk can have an indirect negative influence on technology usage through Perceived Enjoyment.

One specific kind of risk that might be of relevance in the context of smart-glasses is Perceived Health Risk, which we describe as the extent to which a person believes that using smartglasses has negative consequences in terms of his/her health. More specifically, smartglasses can emit radiation; multiple studies have suggested that exposing the human body to radiation might have adverse health consequences such as the development of illnesses (Burgess 2002; Seigneur et al. 2010). For example, the World Health Organization (2013) has warned that radiation might increase the risk of developing cancer, and Myung et al. (2009) has suggested that the risk of developing tumors increases after using devices such as mobile phones. Moreover, popular media has also reported on the potential health risks associated with wearable device usage (e.g., Bilton 2015), which might increase the general public's awareness of the problem.

3 Research Model

In the following section, we will present our research model in Fig. 1 and then outline our corresponding hypotheses.

As described earlier, smartglasses can be used for hedonic purposes such as video games. Therefore, smartglasses are at least partly hedonic technologies (cf. Van der Heijden 2004) that provide positive feelings and experiences for their users in the form of Perceived Enjoyment. Perceived Enjoyment has been shown to be an important antecedent of hedonic technology usage (e.g., Ernst et al. 2013; Van der Heijden 2004). We hypothesize that:

H1   *There is a positive influence of Perceived Enjoyment on the Behavioral Intention to Use*[1] *smartglasses.*

The Theory of Reasoned Action (Fishbein and Ajzen 1975) postulates that an individual's behavior is influenced by his/her particular beliefs concerning the behavior's consequences (e.g., Perceived Enjoyment). Consequently, Perceived Health Risk can be expected to exert an influence on the usage of smartglasses. More precisely, since Perceived Health Risk is associated with negative feelings, the influence it could be exerting is probably negative. Indeed, multiple studies

---

[1]Since at the time of the survey (June 2015), the smartglasses under study, Microsoft HoloLens, were not yet available to the general public, we only included Behavioral Intention to Use, and not Actual System Use, into our research model. Behavioral Intention to Use is a commonly accepted mediator between people's beliefs and their actual behavior. It "capture[s] the motivational factors that influence a [person's] behavior; they are indications of how hard people are willing to try, of how much of an effort they are planning to exert, in order to perform the behavior" (Ajzen 1991, p. 181).

Fig. 1 Research model

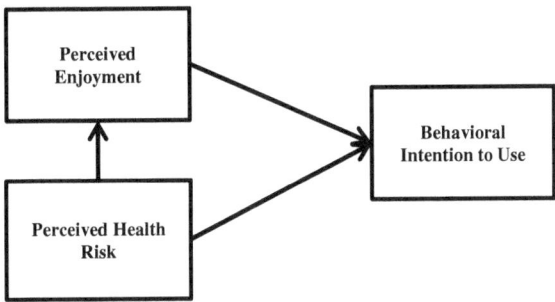

from a variety of contexts have confirmed that various risk perceptions negatively influence technology usage behavior (Featherman and Pavlou 2003; Jarvenpaa et al. 2000; Pavlou 2001, 2003). We hypothesize that:

H2 There is a negative influence of Perceived Health Risk on the Behavioral Intention to Use smartglasses.

Moreover, Perceived Risk, in general, can alter an individual's feelings (Yüksel and Yüksel 2007). More specifically, due to the perceived negative consequences associated with it, Perceived Risk causes negative feelings such as anxiety, discomfort and uncertainty (Dowling and Staelin 1994; Featherman 2001). Indeed, in a shopping context, Yüksel and Yüksel (2007) confirmed a negative influence of Perceived Risk on Pleasure, which is "the degree to which the person feels good, joyful, happy, or satisfied in the situation" (Yüksel and Yüksel 2007, p. 706). In line with this, Ernst (2014) confirmed in a Social Network Site context that Perceived Privacy Risk exerts a negative influence on Perceived Enjoyment. In this sense, due to the potential negative consequences of smartglasses on an individual's health, Perceived Health Risk can also be expected to cause negative feelings, i.e., to negatively influence Perceived Enjoyment. We hypothesize that:

H3 There is a negative influence of Perceived Health Risk on the Perceived Enjoyment of smartglasses.

4 Research Design

4.1 Data Collection

To empirically evaluate our research model, we collected 109 completed German-language online questionnaires about one specific pair of smartglasses, Microsoft HoloLens. At the beginning of the questionnaire, we provided a short description of Microsoft HoloLens, including official images and an explanation of its general functionalities. Microsoft HoloLens, which was not yet available at the time of the survey (June 2015), promises users to see "high-definition holograms … seamlessly integrating with … physical places, spaces, and things" (Microsoft 2015).

56 of our respondents were male (51.38 %) and 53 were female (48.62 %). The average age was 27.58 years (standard deviation: 7.33). 1 respondent was unemployed (0.9 %), 3 were apprentices (2.6 %), 5 were pupils (4.6 %), 26 were currently employed (23.85 %), 12 were self-employed (11.0 %), 61 were students (56.0 %), and 1 selected "other" as a description of themselves (0.9 %).

4.2   Measurement

We adapted existing reflective scales to our context in order to measure Behavioral Intention to Use, Perceived Health Risk, and Perceived Enjoyment. Table 1 presents the resulting reflective items with their corresponding sources. All items were measured using a seven-point Likert-type scale ranging from "strongly agree" to "strongly disagree".

5   Results

Since our data was not distributed joint multivariate normal (cf. Hair et al. 2011), we used the Partial-Least-Squares approach via SmartPLS 3.2.0 (Ringle et al. 2015). With 109 datasets, we met the suggested minimum sample size threshold of

Table 1 Items of our measurement model

Construct	Items	Adapted from
Behavioral Intention to Use	I intend to use a HoloLens in the next 6 months (BI1)	Hu et al. (2011) Venkatesh et al. (2003)
	I predict that I will use a HoloLens in the coming 6 months (BI2)	
	In the future, I am very likely to use a HoloLens (BI3)	
Perceived Enjoyment	Using HoloLens seems to be fun (PE1)	Davis et al. (1992)
	Using HoloLens would be enjoyable (PE2)	
	Using HoloLens would be exciting (PE3)	
Perceived Health Risk	I think using HoloLens would not cause adverse health effects (PHR1) [reversed]	Zhang et al. (2012)
	I believe that the use of HoloLens involves health risks (PHR2)	
	HoloLens involves risks for its user's health (PHR3)	

"ten times the largest number of structural paths directed at a particular latent construct in the structural model" (Hair et al. 2011, p. 144). To test for significance, we used the integrated Bootstrap routine with 5,000 samples (Hair et al. 2011).

In the following section, we will evaluate our measurement model. Indeed, we will examine the indicator reliability, the construct reliability, and the discriminant validity of our reflective constructs. Finally, we will present the results of our structural model.

5.1 Measurement Model

Tables 2 and 3 present the correlations between constructs along with the Average Variance Extracted (AVE) and Composite Reliability (CR), and our reflective items' factor loadings, respectively: All items loaded high (0.783 or more) and significant ($p < 0.001$) on their parent factor and, hence, met the suggested threshold of indicator reliability of 0.70 (Hair et al. 2011); AVE and CR were higher than 0.65 and 0.84, respectively, meeting the suggested construct reliability thresholds of 0.50/0.70 (Hair et al. 2009). The loadings from our reflective indicators were highest for each parent factor and the square root of the AVE of each construct was larger than the absolute value of the construct's correlations with its counterparts, thus indicating discriminant validity (Fornell and Larcker 1981; Hair et al. 2011).

Table 2 Correlations between constructs [AVE (CR) on the diagonal]

	BI	PE	PHR
Behavioral Intention to Use (BI)	0.901 (0.965)		
Perceived Enjoyment (PE)	0.565	0.785 (0.916)	
Perceived Health Risk (PHR)	−0.375	−0.442	0.650 (0.847)

Table 3 Reflective items' loadings (T-Values)

	BI	PE	PHR
BI1	0.934 (44.241)	0.529	−0.367
BI2	0.965 (124.757)	0.568	−0.347
BI3	0.949 (72.825)	0.509	−0.354
PE1	0.493	0.898 (40.020)	−432
PE2	0.547	0.863 (29.383)	−0.363
PE3	0.455	0.896 (35.025)	−0.379
PHR1R	−0.382	−0.455	0.807 (9.773)
PHR2	−0.240	−0.242	0.783 (7.221)
PHR3	−0.234	−0.307	0.827 (8.915)

Fig. 2 Findings

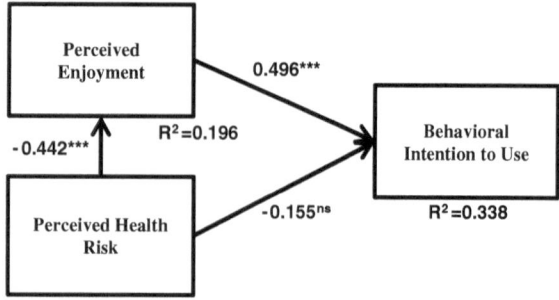

5.2   Structural Model

Figure 2 presents the path coefficients of the previously hypothesized relationships as well as the R^2s of both endogenous variables (*** = $p < 0.001$; ns = non-significant). Hypothesis 2 was not confirmed since Perceived Health Risk, in line with the findings of other similar risk-related studies (e.g., Ernst 2014; Von Stetten et al. 2011), had no significant influence on Behavioral Intention to Use ($\beta = -0.155$, $t = 1.783$). However, it was found to have a significant negative influence on Perceived Enjoyment ($\beta = -0.442$, $p < 0.001$), which, in turn, was found to have a positive influence on Behavioral Intention to Use ($\beta = 0.496$, $p < 0.001$), confirming hypotheses 3 and 1, respectively.

Our research model included two predecessors of Behavioral Intention to Use (Perceived Enjoyment and Perceived Health Risk), and one predecessor of Perceived Enjoyment (Perceived Health Risk). By taking this into account, the explanatory power of our structural model is good, since it explains 33.8 % of the variances of Behavioral Intention to Use as well as 19.6 % of the variances of Perceived Enjoyment.

6   Conclusions

In this article, we evaluated the potential influence of Perceived Health Risk on the usage of smartglasses. After collecting 109 completed online questionnaires about one specific pair of smartglasses, Microsoft HoloLens, and applying a structural equation modeling approach, our findings indicate that smartglasses are at least partly hedonic technologies whose usage is influenced by Perceived Enjoyment. Furthermore, although we could not confirm a direct negative influence of Perceived Health Risk on the Behavioral Intention to Use smartglasses, we confirmed an indirect negative influence of Perceived Health Risk on Behavioral Intention to Use through Perceived Enjoyment.

Our findings have important practical implications. Indeed, they suggest that smartglasses manufacturers need to emphasize the hedonic benefits of their devices as well as address people's potential negative perceptions of the devices in terms of their health. For example, they could collaborate with respected medical experts or ask health-oriented athletes to provide testimonials about the products in order to convince people that their health will not be adversely affected by the use of smartglasses.

Our study has some limitations. First, our empirical findings are based on only one specific pair of smartglasses: Microsoft HoloLens. Hence, the results found for these particular smartglasses might be very different in studies that use other smartglasses. In addition, Microsoft HoloLens is a product that is not yet available to the general public. Hence, our respondents did not have any hands-on experience with the device and could only state their guesses based on our product description as well as on information they might have gathered on their own. Moreover, since we only surveyed German-speaking people, our results might not hold true for non-German speakers. Also, our sample individuals were relatively young (mean: 27.58 years; standard deviation: 7.33). Hence, differences might be found for other age groups. Finally, our survey was only conducted online and, hence, excluded people that do not use the Internet (which might also explain the lack of older people in our sample).

As a next step, we plan to expand our research and address its limitations. More specifically, we want to rollout our survey to a number of other countries around the world and with different smartglasses such as Oculus Rift, in order to evaluate for potential differences between countries and devices. Also, we plan to identify and empirically evaluate a number of additional influence factors of smartglasses usage. For example, the substitutability of physical objects through virtual ones could be an exciting subject to consider and we would like to take a closer look at the implications of this possibility in terms of smartglasses usage.

References

Ajzen, I. (1991). The theory of planned behavior. *Organizational Behavior and Human Decision Processes, 50*(2), 179–211.

Amorim, J. A., Matos, C., Cuperschmid, A. R. M., Gustavsson, P. M., & Pozzer, C. T. (2013). Augmented reality and mixed reality technologies: Enhancing training and mission preparation with simulations. In *STO Modelling and Simulation Group Conference 2013 Proceedings.* Paper 111.

Azuma, R. T. (1997). A survey of augmented reality. *Presence: Teleoperators and Virtual Environments, 6*(4), 355–385.

Bilton, N. (2015). The health concerns in wearable tech. http://www.nytimes.com/2015/03/19/style/could-wearable-computers-be-as-harmful-as-cigarettes.html. Accessed June 16, 2015.

Brief, A. P., & Aldag, R. J. (1977). The intrinsic-extrinsic dichotomy: Toward conceptual clarity. *Academy of Management Review, 2*(3), 496–500.

Burgess, A. (2002). Comparing national responses to perceived health risks from mobile phone masts. *Health, Risk and Society, 4*(2), 175–188.

Chen, R. (2013). Member use of social networking sites—An empirical examination. *Decision Support Systems, 54*(3), 1219–1227.

Davis, F. D., Bagozzi, R. P., & Warshaw, P. R. (1992). Extrinsic and intrinsic motivation to use computers in the workplace. *Journal of Applied Social Psychology, 22*(14), 1111–1132.

Dowling, G. R., & Staelin, R. (1994). A model of perceived risk and intended risk-handling activity. *Journal of Consumer Research, 21*(1), 119–134.

Due, B. L. (2014). The future of smart glasses: An essay about challenges and possibilities with smart glasses. *Working papers on interaction and communication, 1*(2), 1–21 (University of Copenhagen).

Egea, J. M. O., & González, M. V. R. (2010). Explaining physicians' acceptance of EHCR systems: An extension of TAM with trust and risk factors. *Computers in Human Behavior, 27* (1), 319–332.

Ernst, C.-P. H. (2014). Risk hurts fun: The influence of perceived privacy risk on social network site usage. In *AMCIS 2014 Proceedings*.

Ernst, C.-P. H., Pfeiffer, J. & Rothlauf, F. (2013). Hedonic and utilitarian motivations of social network site adoption. Johannes Gutenberg University Mainz, Working Paper.

Featherman, M. (2001). Extending the technology acceptance model by inclusion of perceived risk. In *AMCIS 2001 Proceedings*. Paper 148.

Featherman, M. S., & Pavlou, P. A. (2003). Predicting e-services adoption: A perceived risk facets perspective. *International Journal of Human-Computer Studies, 59*(4), 451–474.

Fishbein, M., & Ajzen, I. (1975). *Belief, attitude, intention, and behavior: An introduction to theory and research*. Reading, MA: Addison-Wesley.

Fornell, C., & Larcker, D. F. (1981). Evaluating structural equation models with unobservable variables and measurement error. *Journal of Marketing Research, 18*(1), 39–50.

Hair, J. F., Black, W. C., Babin, B. J., & Anderson, R. E. (2009). *Multivariate data analysis* (7th ed.). Upper Saddle River, NJ: Prentice Hall.

Hair, J. F., Ringle, C. M., & Sarstedt, M. (2011). PLS-SEM: Indeed a silver bullet. *Journal of Marketing Theory and Practice, 19*(2), 139–151.

Hu, T., Poston, R. S., & Kettinger, W. J. (2011). Nonadopters of online social network services: Is it easy to have fun yet? *Communications of the Association for Information Systems, 29*(1), 441–458.

Jarvenpaa, S. L., Tractinsky, N., & Vitale, M. (2000). Consumer trust in an internet store: A cross-culture validation. *Journal of Computer-Mediated Communication, 5*(2), 45–71.

Kim, M., Francis, A., Gupta, R., & Kumar, M. (2013). Google Glass: Insurance's next killer app. Cognizant White Paper.

Lardinois, F. (2015). Google Glass explorer program shuts down, team now reports to Tony Fadell. http://techcrunch.com/2015/01/15/google-glass-exits-x-labs-as-explorer-program-shuts-down-team-now-reports-to-tony-fadell. Accessed June 29, 2015.

McNaney, R., Vines, J., Roggen, D., Balaam, M., Zhang, P., Poliakov, I., et al. (2014). Exploring the acceptability of Google Glass as an everyday assistive device for people with parkinson's. In *CHI 2014 Proceedings* (pp. 2551–2554).

Microsoft. (2015). Holographic computing is here. http://www.microsoft.com/microsoft-hololens. Accessed July 7, 2015.

Milgram, P., Takemura, H., Utsumi, A., & Kishino, F. (1994). Augmented reality: A class of displays on the reality-virtuality continuum. In *Telemanipulator and Telepresence Technologies 1994 Proceedings* (pp. 282–292).

Myung, S.-K., Ju, W., McDonnell, D. D., Lee, Y. J., Kazinets, G., Cheng, C. T., & Moskowitz, J. M. (2009). Mobile phone use and risk of tumors: A meta-analysis. *Journal of Clinical Oncology, 27*(33), 5565–5572.

Nguyen, T. H. (2013). Innovation insight: Augmented reality will become an important workplace tool. https://www.gartner.com/doc/2640230/innovation-insight-augmented-reality-important. Accessed June 16, 2015.

Nilsson, S., & Johansson, B. (2007). Fun and usable: Augmented reality instructions in a hospital setting. In *Australasian Conference on Computer-Human Interaction 2007 Proceedings* (pp. 123–130).

Pavlou, P. A. (2001). Integrating trust in electronic commerce with the technology acceptance model: Model development and validation. In *AMCIS 2001 Proceedings*. Paper 159.

Pavlou, P. A. (2003). Consumer acceptance of electronic commerce: Integrating trust and risk with the technology acceptance model. *International Journal of Electronic Commerce, 7*(3), 69–103.

Peter, J. P., & Ryan, M. J. (1976). An investigation of perceived risk at the brand level. *Journal of Marketing Research, 13*(2), 184–188.

Rauschnabel, P. A., Brem, A., & Ivens, B. S. (2015). Who will buy smart glasses? Empirical results of two pre-market-entry studies on the role of personality in individual awareness and intended adoption of Google Glass wearables. *Computers in Human Behavior, 49*, 635–647.

Ringle, C. M., Wende, S., & Becker, J.-M. (2015). SmartPLS 3. http://www.smartpls.com.

Seigneur, J.-M., Xavier, T., & Tewfiq, M. (2010). Mobile/wearable device electrosmog reduction through careful network selection. In *AHIC 2010 Proceedings*. Paper 21.

Sitkin, S. B., & Pablo, A. L. (1992). Reconceptualizing the determinants of risk behavior. *Academy of Management Review, 17*(1), 9–38.

Statista. (2014). Shipments of smart glasses worldwide from 2013 to 2015. http://www.statista.com/statistics/302717/smart-glasses-shipments-worldwide. Accessed June 29, 2015.

Tan, S. (1999). Strategies for reducing consumers' risk aversion in internet shopping. *Journal of Consumer Marketing, 16*(2), 163–180.

Tehrani, K. and Andrew, M. (2014): Wearable technology and wearable devices: Everything you need to know. http://www.wearabledevices.com/what-is-a-wearable-device. Accessed June 29, 2015.

Van der Heijden, H. (2004). User acceptance of hedonic information systems. *MIS Quarterly, 28*(4), 695–704.

Venkatesh, V. (2000). Determinants of perceived ease of use: Integrating control, intrinsic motivation, and emotion into the technology acceptance model. *Information Systems Research, 11*(4), 342–365.

Venkatesh, V., Morris, M. G., Davis, G. B., & Davis, F. D. (2003). User acceptance of information technology: Toward a unified view. *MIS Quarterly, 27*(3), 425–478.

Venkatesh, V., Thong, J. Y. L., & Xu, X. (2012). Consumer acceptance and use of information technology: Extending the unified theory of acceptance and use of technology. *MIS Quarterly, 36*(1), 157–178.

Von Stetten, A., Wild, U., & Chrennikow, W. (2011). Adopting social network sites—The role of individual IT culture and privacy concerns. In *AMCIS 2011 Proceedings*. Paper 290.

World Health Organization. (2013). *Non-ionizing radiation, part 2: Radiofrequency electromagnetic fields*. IARC Working Group White Paper.

Yüksel, A., & Yüksel, F. (2007). Shopping risk perceptions: Effects on tourists' emotions, satisfaction and expressed loyalty intentions. *Tourism Management, 28*(3), 703–713.

Zhang, L., Tan, W., Xu, Y., & Tan, G. (2012). Dimensions of consumers' perceived risk and their influences on online consumers' purchasing behavior. *Communications in Information Science and Management Engineering, 2*(7), 8–14.

The Potential Influence of Privacy Risk on Activity Tracker Usage: A Study

Florian Rheingans, Burhan Cikit and Claus-Peter H. Ernst

Abstract Activity trackers collect a broad range of physical activity data and other health-related data. As a result, Perceived Privacy Risk might be a factor hindering people's usage of these devices. In this article, we postulate that Perceived Privacy Risk has both a direct negative influence on the Behavioral Intention to Use activity trackers as well as an indirect influence on the Behavioral Intention to Use them through Perceived Enjoyment. After collecting 115 completed online question- naires and applying a structural equation modeling approach, our findings indicate that activity trackers are at least partly hedonic technologies whose usage is influenced by Perceived Enjoyment. However, we were not able to confirm a significant influence of Perceived Privacy Risk on either the Behavioral Intention to Use the activity trackers or their Perceived Enjoyment. These findings suggest that activity tracker manufacturers need to emphasize the hedonic benefits of their devices and that they do not currently need to address people's potential negative perceptions of activity trackers in terms of privacy risks.

1 Introduction

Wearable devices—i.e., "electronic technologies or computers that are incorporated into items of clothing and accessories which can comfortably be worn on the body" (Tehrani and Andrew 2014)—have gained momentum in the marketplace over the past years. According to (IDC 2015), 26.4 million wearable devices were shipped in 2014 and IDC predicts that by 2019 this number will have grown to 155.7 million per year. Wearable devices come in a variety of forms, from earpieces and watches to belts, glasses and clothes (Poslad 2009). This diversity of products means that wearable devices have a wide range of applications, and they have already been introduced to the fields of health and medicine, fitness, sports and business (PwC 2014).

F. Rheingans · B. Cikit · C.-P.H. Ernst (✉)
Frankfurt University of Applied Sciences, Frankfurt am Main, Germany
e-mail: cernst@fb3.fra-uas.de

© Springer International Publishing Switzerland 2016
C.-P.H. Ernst (ed.), *The Drivers of Wearable Device Usage*,
Progress in IS, DOI 10.1007/978-3-319-30376-5_3

25

One of the most popular forms of wearable devices are activity trackers, which are usually worn on the wrist and provide users with several functions for tracking physical activity data and health-related data such as heartbeat, steps taken, and number of hours of sleep (Miller 2015). Due to this comprehensive collection of sensitive data, the use of activity trackers might carry risks in terms of users' privacy, since users cannot know and/or control how, when, or to what extent, someone might (mis)use the information collected (cf. Westin 1968).

Perceived Risk, in general, can exert an influence on people's behavior (e.g., Tan 1999). Indeed, multiple studies have confirmed the existence of a negative influence of different facets of Perceived Risk on the usage of technologies (e.g., Featherman and Pavlou 2003). In line with this, Perceived Privacy Risk might be a factor hindering people's activity tracker usage. However, to the best of our knowledge, no study has yet empirically evaluated the role of Perceived Privacy Risk on activity tracker usage.

In this article, we postulate that Perceived Privacy Risk has both a direct negative influence on the Behavioral Intention to Use activity trackers and an indirect negative influence on the Behavioral Intention to Use the activity trackers through Perceived Enjoyment, which is a commonly accepted antecedent of hedonic technology usage (e.g., Van der Heijden 2004). After collecting 115 complete online questionnaires about one specific activity tracker, GoBe, and applying a structural equation modeling approach, our findings indicate that activity trackers are at least partly hedonic technologies whose usage is influenced by Perceived Enjoyment. However, we were not able to confirm a significant influence of Perceived Privacy Risk on either Behavioral Intention to Use or Perceived Enjoyment. These findings suggest that activity tracker manufacturers need to emphasize the hedonic benefits of their devices and that they do not currently need to address people's potential negative perceptions of activity trackers in terms of privacy risks.

In the next section, we will present background information on activity trackers, introduce Perceived Enjoyment as an influence factor of hedonic technologies, and also present the theoretical foundations of Perceived Privacy Risk. Following this, we will present our research model and research design. We will then reveal and discuss our results before summarizing our findings, presenting their theoretical and practical implications, and provide an outlook on further research.

2 Theoretical Background

2.1 Activity Trackers

Activity trackers are devices that are typically worn on the body (for example, wristbands) or are attached to shoes, clothes, or other wearable accessories. They usually contain multiple sensors (for example, accelerometers and gyroscopic sensors) that allow them to track physical activity data and health-related data such as heartbeat, steps taken, and number of hours of sleep. The analyses functions of this data are usually done on separate, more powerful devices such as smartphones or PCs (Barcena et al. 2014; Miller 2015).

Although multiple studies have studied different aspects of wearable devices (e.g., Ariyatum et al. 2005; Bodine and Gemperle 2003; Dvorak 2008; Starner 2001), the factors that drive peoples' activity tracker usage are largely unknown. Indeed, to the best of our knowledge, there is only one article that has studied factors driving activity tracker usage. This recent empirical study suggests that product design, the ability to measure one's heart rate, and product quality are important drivers of activity tracker's usage (Seiler and Hüttermann 2015).

2.2 The Role of Perceived Enjoyment on Activity Tracker Usage

(Seiler and Hüttermann's 2015) findings suggest that activity tracker usage positively influences the practice of physical activities. More specifically, activity tracker usage might have a positive effect on training regularity, performance improvement, and training efficiency.

Exercise is often seen as a leisure activity and is generally accepted to provide people with hedonic benefits such as enjoyment, fun, etc. (e.g., Côté and Hay 2002; MacPhail et al. 2003; Nielsen et al.2014; Thedin Jakobsson 2014; Vlachopoulos et al. 2000). In line with the findings of (Seiler and Hüttermann 2015), it is highly probable that activity tracker usage is associated with these hedonic contexts, making activity trackers at least a partly hedonic technology.

Generally, hedonic technologies "aim to provide self-fulfilling value to the user, … [which] is a function of the degree to which the user experiences fun when using the system" (Van der Heijden 2004, p. 696). Various studies in multiple contexts have consistently confirmed that Perceived Enjoyment—"the extent to which the activity of using a specific system is perceived to be enjoyable in its own right, aside from any performance consequences resulting from system use" (Venkatesh 2000, p. 351)—is a central antecedent of hedonic technologies' usage (e.g.,Van der Heijden 2004). By applying these findings to our contexts, a person can be expected to use activity trackers if he/she believes that they fulfill his/her expectations with regards to enjoyment.

2.3 Perceived Privacy Risk

Risk can be generally described as "the extent to which there is an uncertainty in significant and disappointing outcomes that may be realized" (Chen 2013, p. 1222; Sitkin and Pablo 1992). Perceived Risk is thus consistently understood as "the expectation of losses associated with ... [specific actions]" (Peter and Ryan 1976, p. 185). Several studies have confirmed that Perceived Risk (e.g., Privacy Risk) can exert a negative influence on the usage of technologies (e.g., Egea and Gonzáles 2010; Tan 1999).

One specific kind of risk that might be of relevance in the context of activity trackers is Perceived Privacy Risk, which can be described as the extent to which a person believes that using an activity tracker has negative consequences with regards to his/her privacy (Ernst 2014; cf. Chen 2013; Dinev and Hart 2006; Featherman and Pavlou 2003; Kim et al. 2008; Krasnova et al. 2010; Peter and Ryan 1976; Wu et al. 2009).

More specifically, since activity trackers collect and store sensitive data on the devices themselves and regularly also on connected devices such as smartphones as well as in the cloud (Barcena et al. 2014), users' privacy—"the claim of individuals ... to determine for themselves when, how, and to what extent information about them is communicated to others" (Westin 1968, p. 7)—could be endangered. In fact, users do not have any control about what their device's manufacturer might do with their data. Moreover, during the transmission from the activity tracker to connected devices as well as during the transmission to the cloud, data could be intercepted by a third party. Furthermore, third parties might gain access to the data stored in the activity tracker, the connected devices, or in the cloud (Barcena et al. 2014). In addition, users might willingly or unwillingly share their sensitive data from the activity trackers or accompanying apps to social media themselves, making them visible to potentially everyone. Resulting negative consequences can be, for example, the discrimination of companies such as health insurance companies due to a person's individual characteristics, or embarrassment due to private information (for example, calorie intake) becoming public (cf. Barcena et al. 2014).

In line with this, Perceived Privacy Risk might be a factor hindering people's usage of activity trackers. Still, to the best of our knowledge, no study has yet empirically evaluated the role of Perceived Privacy Risk on activity tracker usage.

3   Research Model

In the following section, we will present our research model in Fig. 1 and then outline our corresponding hypotheses.

As described earlier, activity trackers are regularly used in hedonic contexts. Therefore, activity trackers can be seen as at least partly hedonic technologies that provide positive feelings and experiences for their users in the form of Perceived Enjoyment (Van der Heijden 2004). Perceived Enjoyment has been shown to be an important antecedent of hedonic technologies' usage (e.g., Ernst et al. 2013; Van der Heijden 2004). We hypothesize that:

Fig. 1 Research model

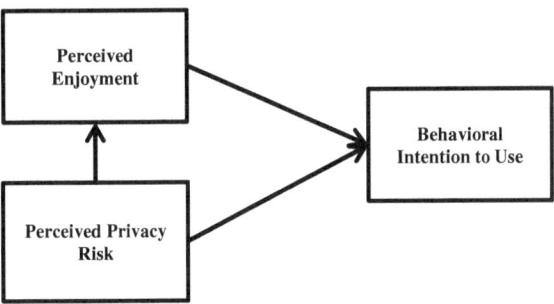

H1   There is a positive influence of Perceived Enjoyment on the Behavioral Intention to Use[1] activity trackers.

The Theory of Reasoned Action (Fishbein and Ajzen 1975) postulates that an individual's behavior is influenced by his/her particular beliefs concerning the behavior's consequences (e.g., Perceived Enjoyment). Consequently, Perceived Privacy Risk can be expected to exert an influence on activity tracker usage. More precisely, since Perceived Privacy Risk is associated with negative feelings, the influence it could be exerting is probably negative. Indeed, multiple studies from different contexts have confirmed that various risk perceptions negatively influence technology usage behavior (Featherman and Pavlou 2003; Jarvenpaa et al. 2000; Pavlou 2001, 2003). We hypothesize that:

H2   There is a negative influence of Perceived Privacy Risk on the Behavioral Intention to Use activity trackers.

Moreover, Perceived Risk, in general, can alter an individual's feelings (Yüksel and Yüksel 2007). More specifically, due to the perceived negative consequences associated with it, Perceived Risk causes negative feelings such as anxiety, discomfort and uncertainty (Dowling and Staelin 1994; Featherman 2001). Indeed, in a shopping context, (Yüksel and Yüksel 2007) confirmed a negative influence of Perceived Risk on Pleasure, which is "the degree to which the person feels good, joyful, happy, or satisfied in the situation" (Yüksel and Yüksel 2007, p. 706). In line with this, (Ernst 2014) confirmed in a Social Network Site context that Perceived

---

[1]Since at the time of this study (June 2015), the activity tracker under study, GoBe, was not yet available to the public in Germany, we only included Behavioral Intention to Use, and not Actual System Use, into our research model. Behavioral Intention to Use is a commonly accepted mediator between people's beliefs and their actual behavior. It "capture[s] the motivational factors that influence a [person's] behavior; they are indications of how hard people are willing to try, of how much of an effort they are planning to exert, in order to perform the behavior" (Ajzen 1991, p. 181).

Privacy Risk exerts a negative influence on Perceived Enjoyment. In this sense, due to the potential negative consequences of activity trackers with regards to an individual's privacy, Perceived Privacy Risk can also be expected to cause negative feelings, i.e., to negatively influence an individual's Perceived Enjoyment. We hypothesize that:

H3 There is a negative influence of Perceived Privacy Risk on activity trackers' Perceived Enjoyment.

4 Research Design

4.1 Data Collection

To empirically evaluate our research model, we collected 115 completed German-language online questionnaires about one specific activity tracker: GoBe. At the beginning of the questionnaire, we gave a short description of GoBe, including official images and an explanation of its general functionalities. GoBe, which was not yet available in Germany at the time of the survey (June 2015), promised users it could track multiple activity-related data and health-related data such as heart rate, blood pressure, stress level, hours of sleep, calorie intake and calories burned (Rubin et al. 2015).

53 of our respondents were male (46.09 %) and 62 were female (53.91 %). The average age was 25.93 years (standard deviation: 5.18). 5 respondents were apprentices (4.3 %), 34 were currently employed (29.6 %), 70 were students (60.9 %), and 6 selected "other" as a description of themselves (5.2 %).

4.2 Measurement

We adapted existing reflective scales to our context in order to measure the Behavioral Intention to Use the activity tracker and its Perceived Enjoyment. For Perceived Privacy Risk, we adapted three items from (Chen 2013), (Featherman and Pavlou 2003), and (Krasnova et al. 2010). Table 1 presents the resulting reflective items with their corresponding sources. All items were measured using a seven-point Likert-type scale ranging from "strongly agree" to "strongly disagree".

Table 1 Items of our measurement model

Construct	Items (Labels)	Source/adapted from
Behavioral Intention to Use	I intend to use GoBe in the next 6 months (BI1)	Hu et al. (2011) Venkatesh et al. (2003)
	In the future, I am very likely to use GoBe (BI2)	
	I predict that i will use GoBe in the next 6 months (BI3)	
Perceived Enjoyment	Using GoBe would be exciting (PE1)	Davis et al. (1992)
	Using GoBe would be enjoyable (PE2)	
	Using GoBe would be fun (PE3)	
Perceived Privacy Risk	Using GoBe would allows others to misuse my personal data (PPR1)	Chen (2013) Featherman and Pavlou (2003) Krasnova et al. (2010)
	I think that GoBe users' privacy is at risk (PPR2)	
	Overall, I see a privacy threat linked to GoBe's usage (PPR3)	

5 Results

Since our data was not distributed joint multivariate normal (cf. Hair et al. 2011), we used the Partial-Least-Squares approach via SmartPLS 3.2.0 (Ringle et al. 2015). With 115 datasets, we met the suggested minimum sample size threshold of "ten times the largest number of structural paths directed at a particular latent construct in the structural model" (Hair et al. 2011, p. 144). To test for significance, we used the integrated Bootstrap routine with 5,000 samples (Hair et al. 2011).

In the following section, we will evaluate our measurement model. Indeed, we will examine the indicator reliability, construct reliability, and discriminant validity of our reflective constructs. Finally, we will present the results of our structural model.

5.1 Measurement Model

Tables 2 and 3 present the correlations between constructs along with the Average Variance Extracted (AVE) and Composite Reliability (CR), and our reflective items' factor loadings, respectively: All items loaded high (0.791 or more) and significant ($p < 0.001$) on their parent factor and, hence, met the suggested threshold of indicator reliability of 0.70 (Hair et al. 2011); AVE and CR were higher than 0.76 and 0.90, respectively, meeting the suggested construct reliability thresholds of 0.50/0.70 (Hair et al. 2009). The loadings from our reflective indicators were highest for each parent factor and the square root of the AVE of each construct was larger than the absolute value of the construct's correlations with its counterparts, thus indicating discriminant validity (Fornell and Larcker 1981; Hair et al. 2011).

Table 2 Correlations between constructs [AVE (CR) on the diagonal]

	BI	PE	PPR
Behavioral Intention to Use (BI)	0.936 (0.978)		
Perceived Enjoyment (PE)	0.535	0.767 (0.908)	
Perceived Privacy Risk (PPR)	−0.058	−0.135	0.837 (0.939)

Table 3 Reflective items' loadings (T-Values)

	BI	PE	PPR
BI1	0.965 (114.998)	0.490	−0.078
BI2	0.968 (90.841)	0.552	−0.062
BI3	0.969 (147.816)	0.508	−0.027
PE1	0.373	0.791 (16.712)	−0.047
PE2	0.488	0.926 (52.165)	−0.068
PE3	0.524	0.904 (16.836)	−0.212
PPR1	−0.013	−0.093	0.913 (6.851)
PPR2	−0.083	−0.161	0.976 (7.030)
PPR3	−0.006	−0.029	0.851 (5.880)

5.2 Structural Model

Figure 2 presents the path coefficients of the previously hypothesized relationships as well as the R^2s of both endogenous variables (*** = $p < 0.001$; ns = non-significant). Perceived Enjoyment was found to have a positive influence on Behavioral Intention to Use ($\beta = 0.537$, $p < 0.001$), confirming hypothesis 1. However, hypotheses 2 and 3 could not be confirmed since Perceived Privacy Risk had no significant influence on either Behavioral Intention to Use ($\beta = 0.015$, $t = 0.266$; cf. Ernst 2014; Von Stetten et al. 2011) or Perceived Enjoyment ($\beta = −0.135$, $t = 1.684$).

Fig. 2 Findings

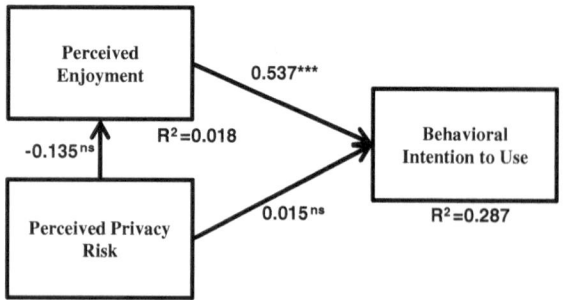

In summary, our findings indicate that activity trackers are at least partly hedonic technologies whose usage is influenced by Perceived Enjoyment. Moreover, Privacy Risks seem to play no part in people's usage of these devices.

6   Conclusions

In this article, we evaluated the potential influence of Perceived Privacy Risk on activity tracker usage. After collecting 115 complete online questionnaires and applying a structural equation modeling approach, our findings indicate that activity trackers are at least partly hedonic technologies whose usage is influenced by Perceived Enjoyment. However, we were not able to confirm a significant influence of Perceived Privacy Risk on either the Behavioral Intention to Use the activity tracker or its Perceived Enjoyment.

Our findings have important practical implications. Indeed, they suggest that activity tracker manufacturers currently do not need to address people's potential negative perceptions with regards to their privacy and can focus on other matters, such as emphasizing the hedonic benefits of their devices.

Our study has some limitations. First, our empirical findings are based on only one specific activity tracker: GoBe. Therefore, there might be differences between this particular activity tracker and other activity trackers. Additionally, GoBe was not yet available to the public in Germany at the time of this study (June 2015). Hence, our respondents did not have any hands-on experience with the device and could only state their guesses based on our product description as well as on information they might have gathered on their own. Moreover, since we only surveyed German-speaking people, our results might not hold true for non-German speaking people. Also, our sample individuals were relatively young (mean: 25.93 years; standard deviation: 5.18). Hence, differences might be found for other age groups. Finally, our survey was only conducted online and, hence, excluded people that do not use the Internet (which might also explain the lack of older people in our sample).

As a next step, we plan to expand our research and address its limitations. More specifically, we want to rollout our survey to a greater number of countries around the world, focusing on different activity trackers as well as on smartwatches (which regularly incorporate the functionalities of activity trackers but also provide additional functionalities) that are already on the market, in order to evaluate the potential differences between countries and devices. Also, we plan to identify and empirically evaluate additional influence factors of activity tracker usage. For example, whereas technology research has often studied the potential negative effects of technology usage on people's health such as cancer through radiation, activity trackers might actually exert a positive effect on individuals by motivating them to exercise more. Hence, we plan to study the implications of Perceived Health Increase on activity tracker usage.

References

Ajzen, I. (1991). The theory of planned behavior. *Organizational Behavior and Human Decision Processes, 50*(2), 179–211.

Ariyatum, B., Holland, R., Harrison, D., & Kazi, T. (2005). The future design direction of smart clothing development. *Journal of the Textile Institute, 96*(4), 199–210.

Barcena, M. B., Wueest, C., & Lau, H. (2014). *How safe is your quantified self?* Symantec White Paper.

Bodine, K., & Gemperle, F. (2003). Effects of functionality on perceived comfort of wearables. In *IEEE International Symposium on Wearable Computers 2003 Proceedings.*

Chen, R. (2013). Member use of social networking sites—An empirical examination. *Decision Support Systems, 54*(3), 1219–1227.

Côté, J., & Hay, J. (2002). Children's involvement in sport: A development perspective. In J. Silva & D. Stevens (Eds.), *Psychological Foundations of Sport* (pp. 484–502). Boston, MA: Allyn & Bacon.

Davis, F. D., Bagozzi, R. P., & Warshaw, P. R. (1992). Extrinsic and intrinsic motivation to use computers in the workplace. *Journal of Applied Social Psychology, 22*(14), 1111–1132.

Dinev, T., & Hart, P. (2006). An extended privacy calculus model for e-commerce transactions. *Information Systems Research, 17*(1), 61–80.

Dowling, G. R., & Staelin, R. (1994). A model of perceived risk and intended risk handling activity. *Journal of Consumer Research, 21*(1), 119–134.

Dvorak, J. L. (2008). *Moving wearables into the mainstream: Taming the Borg.* New York: Springer.

Egea, J. M. O., & González, M. V. R. (2010). Explaining physicians' acceptance of EHCR systems: An extension of TAM with trust and risk factors. *Computers in Human Behavior, 27*(1), 319–332.

Ernst, C.-P. H. (2014). Risk hurts fun: The influence of perceived privacy risk on social network site usage. In *AMCIS 2014 Proceedings.*

Ernst, C.-P. H., Pfeiffer, J., & Rothlauf, F. (2013). *Hedonic and utilitarian motivations of social network site adoption.* Johannes Gutenberg University Mainz, Working papers.

Featherman, M. (2001). Extending the technology acceptance model by inclusion of perceived risk. In *AMCIS 2001 Proceedings.* Paper 148.

Featherman, M. S., & Pavlou, P. A. (2003). Predicting e-services adoption: A perceived risk facets perspective. *International Journal of Human-Computer Studies, 59*(4), 451–474.

Fishbein, M., & Ajzen, I. (1975). *Belief, attitude, intention, and behavior: An introduction to theory and research.* Reading, MA: Addison-Wesley.

Fornell, C., & Larcker, D. F. (1981). Evaluating structural equation models with unobservable variables and measurement error. *Journal of Marketing Research, 18*(1), 39–50.

Hair, J. F., Black, W. C., Babin, B. J., & Anderson, R. E. (2009). *Multivariate data analysis* (7th ed.). Upper Saddle River, NJ: Prentice Hall.

Hair, J. F., Ringle, C. M., & Sarstedt, M. (2011). PLS-SEM: Indeed a silver bullet. *Journal of Marketing Theory and Practice, 19*(2), 139–152.

Hu, T., Poston, R. S., & Kettinger, W. J. (2011). Nonadopters of online social network services: Is it easy to have fun yet? *Communications of the Association for Information Systems, 29*(1), 441–458.

IDC (2015). Worldwide wearables market forecast to grow 173.3 % in 2015 with 72.1 million units to be shipped, according to IDC. http://www.idc.com/getdoc.jsp?containerId=prUS25696715. Accessed May 22, 2015.

Jarvenpaa, S. L., Tractinsky, N., & Vitale, M. (2000). Consumer trust in an Internet store. *Information Technology and Management, 1*(1), 45–71.

Kim, D. J., Ferrin, D. L., & Rao, H. R. (2008). A trust-based consumer decision-making model in electronic commerce: The role of trust, perceived risk, and their antecedents. *Decision Support Systems, 44*(2), 544–564.

Krasnova, H., Spiekermann, S., Koroleva, K., & Hildebrand, T. (2010). Online social networks: Why we disclose. *Journal of Information Technology, 25*(2), 109–125.

MacPhail, A., Gorely, T., & Kirk, D. (2003). Young people's socialisation into sport: A case study of an athletics club. *Sport, Education and Society, 8*(2), 251–267.

Miller, M. (2015). *The Internet of things: How smart TVs, smart cars, smart homes, and smart cities are changing the world.* Indianapolis, IN: Que.

Nielsen, G., Wikman, J. M., Jensen, C. J., Schmidt, J. F., Gliemann, L., & Andersen, T. R. (2014). Health promotion: The impact of beliefs of health benefits, social relations and enjoyment on exercise continuation. *Scandinavian Journal of Medicine and Science in Sports, 24*(1), 66–75.

Pavlou, P. (2001). Integrating trust in electronic commerce with the technology acceptance model: Model development and validation. In *AMCIS 2001 Proceedings*.

Pavlou, P. A. (2003). Consumer acceptance of electronic commerce: Integrating trust and risk with the technology acceptance model. *International Journal of Electronic Commerce, 7*(3), 101–134.

Peter, J. P., & Ryan, M. J. (1976). An investigation of perceived risk at the brand level. *Journal of Marketing Research, 13*(2), 184–188.

Poslad, S. (2009). *Ubiquitous computing: Smart devices environments and interactions.* Chichester, UK: Wiley.

PwC (2014). Health wearables: Early days. White paper.

Ringle, C. M., Wende, S., & Becker, J. M. (2015) SmartPLS 3. http://www.smartpls.com.

Rubin, M. S., Sokolov, Y. L., & Misyuchenko, I. L. (2015) The history and technology behind Healbe GoBe. Healbe White Paper.

Seiler, R., & Hüttermann, M. (2015). E-Health, fitness trackers and wearables—Use among Swiss students. In *Advances in Business-Related Scientific Research Conference 2015 Proceedings*.

Sitkin, S. B., & Pablo, A. L. (1992). Reconceptualizing the determinants of risk behavior. *Academy of Management Review, 17*(1), 9–38.

Starner, T. (2001). The challenges of wearable computing: Part 2. *IEEE Micro, 21*(4), 54–67.

Tan, S. J. (1999). Strategies for reducing consumers' risk aversion in Internet shopping. *Journal of Consumer Marketing, 16*(2), 163–180.

Tehrani, K., & Michael, A. (2014). Wearable technology and wearable devices: Everything you need to know. http://www.wearabledevices.com/what-is-a-wearable-device. Accessed June 29, 2015.

Thedin Jakobsson, B. (2014). What makes teenagers continue? A salutogenic approach to understanding youth participation in Swedish club sports. *Physical Education and Sport Pedagogy, 19*(3), 239–252.

Van der Heijden, H. (2004). User acceptance of hedonic information systems. *MIS Quarterly, 28*(4), 695–704.

Venkatesh, V. (2000). Determinants of perceived ease of use: Integrating control, intrinsic motivation, and emotion into the technology acceptance model. *Information Systems Research, 11*(4), 342–365.

Venkatesh, V., Morris, M. G., Davis, G. B., & Davis, F. D. (2003). User acceptance of information technology: Toward a unified view. *MIS Quarterly, 27*(3), 425–478.

Vlachopoulos, S. P., Karageorghis, C. I., & Terry, P. C. (2000). Motivation profiles in sport: A self-determination theory perspective. *Research Quarterly for Exercise and Sport, 71*(4), 387–397.

Von Stetten, A., Wild, U., & Chrennikow, W. (2011). Adopting social network sites—The role of individual IT culture and privacy concerns. In *AMCIS 2011 Proceedings*. Paper 290.

Westin, A. F. (1968). *Privacy and freedom.* New York, NY: Atheneum.

Wu, Y. (2009). Influence of social context and affect on individuals' implementation of information security safeguards. In *ICIS 2009 Proceedings*. Paper 70.

Yüksel, A., & Yüksel, F. (2007). Shopping risk perceptions: Effects on tourists' emotions, satisfaction and expressed loyalty intention. *Tourism Management, 28*(3), 703–713.

An Analysis of the Potential Influence of Privacy Risk on Neuroheadset Usage

Frederik M. Mayer, Duc T. Nguyen and Claus-Peter H. Ernst

Abstract Neuroheadsets use electroencephalography (EEG) to record cognitive activity and some neuroheadsets are even capable of deciphering basic mental commands. As a result, users might believe there are privacy risks associated with these devices, which can hinder their usage. In this article, we postulate that Perceived Privacy Risk has both a direct negative influence on the Behavioral Intention to Use neuroheadsets and an indirect negative influence on the Behavioral Intention to Use neuroheadsets through Perceived Usefulness. After collecting 107 completed online questionnaires and applying a structural equation modeling approach, our findings indicate that neuroheadsets are at least partly utilitarian technologies whose usage is influenced by Perceived Usefulness. However, we were not able to confirm a significant influence of Perceived Privacy Risk on either the Behavioral Intention to Use neuroheadsets or their Perceived Usefulness. These findings suggest that neuroheadset manufacturers need to emphasize the instrumental benefits of their devices, but that they do not currently need to address people's potential negative perceptions of neuroheadsets in terms of privacy risks.

1 Introduction

Wearable devices—i.e., "electronic technologies or computers that are incorporated into items of clothing and accessories which can comfortably be worn on the body" (Tehrani and Andrew 2014)—have gained momentum in the marketplace over the past years. According to forecasts, the revenues of wearable devices are expected to exceed 9 billion Euros in 2018 in Europe alone (Statista 2014). Wearable devices come in a variety of forms, from earpieces and watches to belts, glasses and clothes

F.M. Mayer · D.T. Nguyen · C.-P.H. Ernst (✉)
Frankfurt University of Applied Sciences, Frankfurt am Main, Germany
e-mail: cernst@fb3.fra-uas.de

© Springer International Publishing Switzerland 2016   37
C.-P.H. Ernst (ed.), *The Drivers of Wearable Device Usage*,
Progress in IS, DOI 10.1007/978-3-319-30376-5_4

(Poslad 2009). This diversity of products means that wearable devices have a wide range of applications, in the fields of health and medicine, fitness, sports and business (PwC 2014).

One of the most popular forms of wearable devices are physical activity trackers, which are usually worn on the wrist and provide users with several functions for tracking physical activity such as heartbeat, steps taken, and number of hours of sleep (Miller 2015). Another kind of activity tracker are neuroheadsets that use electroencephalography (EEG) to record cognitive activity and promise users improvement in the areas of mental performance, cognitive health, and well-being (cf. Emotiv 2015).

Some neuroheadsets are even capable of deciphering basic mental commands, enabling users to do things such as operating PCs (e.g., Emotiv 2015). Hence, people might have the impression that neuroheadsets can monitor, analyze and interpret their most private details, i.e., their personal thoughts. As a result, users might perceive privacy risks with regards to the devices' usage.

Perceived Risk, in general, can exert an influence on individuals' behavior (e.g., Tan 1999). Indeed, multiple studies have confirmed the existence of a negative influence of different facets of Perceived Risk on the usage of technologies (e.g., Featherman and Pavlou 2003). In line with this, Perceived Privacy Risk could be a factor that would hinder an individual's neuroheadset usage. Still, to the best of our knowledge, no study has yet empirically evaluated the role of Perceived Privacy Risk on neuroheadset usage.

In this article, we postulate that Perceived Privacy Risk has both a direct negative influence on the Behavioral Intention to Use neuroheadsets and an indirect negative influence on the Behavioral Intention to Use the headsets through Perceived Usefulness, which is a commonly accepted antecedent of utilitarian technology usage (e.g., Davis et al. 1989). After collecting 107 complete online questionnaires about one specific neuroheadset, Emotiv Insight, and applying a structural equation modeling approach, our findings indicate that neuroheadsets are at least partly utilitarian technologies whose usage is influenced by Perceived Usefulness. However, we were not able to confirm a significant influence of Perceived Privacy Risk on either Behavioral Intention to Use or Perceived Usefulness. These findings suggest that neuroheadset manufacturers need to emphasize the instrumental benefits of their devices and that they do not currently need to address people's potential negative perceptions of neuroheadsets in terms of privacy risks.

In the next section, we will present background information on neuroheadsets, introduce Perceived Usefulness as an influence factor of utilitarian technologies, and also present the theoretical foundations of Perceived Privacy Risk. Following this, we will present our research model and research design. We will then reveal and discuss our results before summarizing our findings, presenting their theoretical and practical implications, and provide an outlook on further research.

2 Theoretical Background

2.1 Neuroheadsets

Neuroheadsets are devices that are worn on the head. They usually contain multiple sensors (such as an electroencephalography [EEG] sensor, a gyroscope, an accelerometer, and a magnetometer), which allow them to record the cognitive activity of the brain. The neuroheadsets can provide multiple instrumental benefits to users. For example, the headsets can enable users to improve their mental performance, cognitive health, and well-being (cf. Emotiv 2015), and are even capable of deciphering a user's mental commands so that they can "interact with the environment [such as operating PCs, by thought] without the need for muscular or peripheral neural activity" (Lal et al. 2005, p. 1). With these multiple functions, neuroheadsets can be considered, at least in part, as utilitarian technologies.

Although multiple studies have studied different aspects of neuroheadsets [including their usage for research and medical diagnoses (e.g., Vala and Trivedi 2014)], the factors that drive private users' neuroheadset usage are unknown. Indeed, to the best of our knowledge, no study has yet evaluated the influence factors of private users' neuroheadset usage.

2.2 The Role of Perceived Usefulness on Neuroheadset Usage

Generally, utilitarian technologies "aim to provide instrumental value to the user" (Van der Heijden 2004, p. 696). Perceived Usefulness—i.e., "the degree to which a person believes that using a particular system would enhance his or her job [and task] performance" (Davis 1989, p. 320)—centers on the motivations and benefits that are external to the system-user interaction itself. These are referred to as extrinsic motivations (Brief and Aldag 1977; Van der Heijden 2004). For example, the external benefits/extrinsic motivations of a text-processing program can be to foster a good writing performance in terms of a well-structured and orthographically error-free text (Davis et al. 1989).

Various studies in multiple contexts have consistently confirmed that Perceived Usefulness is a central antecedent of utilitarian technologies' usage (e.g., Davis 1989). By applying these findings to our context, a person can be expected to use neuroheadsets if he/she believes that they fulfill his/her expectations with regards to their instrumental benefits, i.e., their Perceived Usefulness.

2.3 Perceived Privacy Risk

Risk can be generally described as "the extent to which there is an uncertainty in significant and disappointing outcomes that may be realized" (Chen 2013, p. 1222; Sitkin and Pablo 1992). Perceived Risk is thus consistently understood as "the expectation of losses associated with … [specific actions]" (Peter and Ryan 1976, p. 185). Several studies have confirmed that Perceived Risk can exert a negative influence on the usage of technologies (e.g., Egea and Gonzáles 2010; Tan 1999).

One specific kind of risk that might be of relevance in the context of neuroheadsets is Perceived Privacy Risk, which can be described as the extent to which a person believes that using a neuroheadset has negative consequences with regards to his/her privacy (Ernst 2014; cf. Chen 2013; Dinev and Hart 2006; Featherman and Pavlou 2003; Kim et al. 2008; Krasnova et al. 2010; Peter and Ryan 1976; Wu et al. 2009). More specifically, some neuroheadsets are capable of deciphering basic mental commands (e.g., Emotiv 2015). Although they are currently (as of July 2015) not capable of actually reading peoples' minds, people might nevertheless believe that this is the case, or they might believe that future technological innovations will enable third parties to retrospectively retrieve their innermost thoughts, passwords, bank data, etc. Hence, people might perceive risks with regards to their privacy, since they cannot know and/or control how, when, or to what extent, someone might (mis)use their information (cf. Westin 1968). In line with this, Perceived Privacy Risk might be a factor hindering people's neuroheadset usage.

3 Research Model

In the following section, we will present our research model in Fig. 1 and then outline our corresponding hypotheses.

As described earlier, neuroheadsets provide instrumental benefits to its users such as enabling them to improve their mental performance, cognitive health, and well-being (cf. Emotiv 2015). Therefore, neuroheadsets are at least partly utilitarian technologies (cf. Ernst et al. 2013) that provide users with benefits that are external to the system-user interaction itself. Perceived Usefulness is commonly accepted to be an important antecedent of utilitarian technologies' usage (e.g., Davis et al. 1989). We hypothesize that:

H1 *There is a positive influence of Perceived Usefulness on the Behavioral Intention to Use[1] neuroheadsets.*

---

[1]Since at the time of the survey (June 2015), the neuroheadset under study, Emotiv Insight, was not yet released to the general public, we only included Behavioral Intention to Use, and not Actual System Use, into our research model. Behavioral Intention to Use is a commonly accepted mediator between people's beliefs and their actual behavior. It "capture[s] the motivational factors that influence a [person's] behavior; they are indications of how hard people are willing to try, of how much of an effort they are planning to exert, in order to perform the behavior" (Ajzen 1991, p. 181).

Fig. 1 Research model

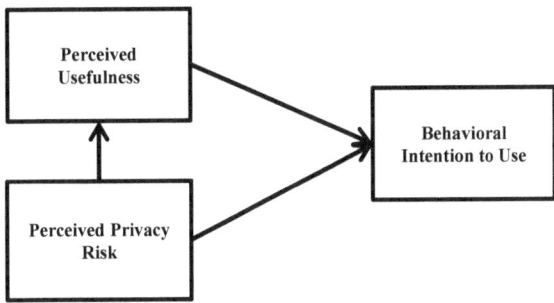

The Theory of Reasoned Action (Fishbein and Ajzen 1975) postulates that an individual's behavior is influenced by his/her particular beliefs concerning the behavior's consequences (e.g., Perceived Usefulness). Consequently, Perceived Privacy Risk can be expected to exert an influence on neuroheadset usage. More precisely, since Perceived Privacy Risk is associated with negative feelings, the influence it could be exerting is probably negative. Indeed, multiple studies from a variety of contexts have confirmed that various risk perceptions negatively influence technology usage behavior (Featherman and Pavlou 2003; Jarvenpaa et al. 2000; Pavlou 2001, 2003). We hypothesize that:

H2 *There is a negative influence of Perceived Privacy Risk on the Behavioral Intention to Use neuroheadsets.*

Moreover, Perceived Risk, in general, causes negative feelings such as anxiety, discomfort and uncertainty (Dowling and Staelin 1994; Featherman 2001) and, hence, inhibits product evaluation (Dowling and Staelin 1994; cf. Featherman and Pavlou 2003). For example, it has been shown that Perceived Risk negatively influences peoples' product evaluations regarding enjoyment (e.g., Ernst 2014) and usefulness (e.g., Rose and Fogarty 2006). In this sense, due to its potential negative consequences with regards to an individual's privacy, Perceived Privacy Risk can be expected to negatively influence an individual's neuroheadset evaluation with regards to its instrumental benefits, i.e., Perceived Usefulness. Thus, we hypothesize that:

H3 *There is a negative influence of Perceived Privacy Risk on neuroheadsets' Perceived Usefulness.*

4 Research Design

4.1 Data Collection

To empirically evaluate our research model, we collected 107 completed English-language online questionnaires about one specific neuroheadset: Emotiv Insight. At the beginning of the questionnaire, we gave a short description of Emotiv Insight, including official images of the headset and an explanation of its general functionalities. Emotiv Insight, which was not yet available to the general public at the time of the survey (June 2015), promised multiple instrumental benefits such as recording and analyzing users' brain's activity in order to measure and track their attention, engagement, and stress levels, etc., thus enabling users to improve their mental performance, cognitive health and well-being, as well as deciphering basic mental commands (Emotiv 2015).

52 of our respondents were male (48.60 %) and 55 were female (51.40 %). The average age was 26.10 years (standard deviation: 6.94). 5 respondents were unemployed (4.7 %), 38 were currently employed (35.5 %), 63 were students (58.9 %), and 1 selected "other" as a description of themselves (0.9 %).

4.2 Measurement

We adapted existing reflective scales to our context in order to measure the Behavioral Intention to Use the neuroheadset and its Perceived Usefulness. For Perceived Privacy Risk, we adapted three items from Chen (2013), Featherman and Pavlou (2003), and Krasnova et al. (2010). Table 1 presents the resulting reflective

Table 1 Items of our measurement model

Construct	Items (labels)	Source/adapted from
Behavioral Intention to Use	I intend to use Emotiv Insight in the next 6 months (BI1)	Hu et al. (2011) Venkatesh et al. (2003)
	In the near future, I will probably use Emotiv Insight (BI2)	
	I think that I will use Emotiv Insight in the next 6 months (BI3)	
Perceived Privacy Risk	I consider that the privacy of Emotiv Insight users' thoughts is at risk (PPR1)	Chen (2013) Featherman and Pavlou (2003) Krasnova et al. (2010)
	Using Emotiv Insight would lead to a loss of control over the privacy of my thoughts (PPR2)	
	Overall, I would see a threat to the privacy of my thoughts due to the usage of Emotiv Insight (PPR3)	
Perceived Usefulness	Overall, Emotiv Insight is useful (PU1)	Alarcón-del-Amo et al. (2012) cf. Ernst et al. (2013)
	The Emotiv Insight is an effective tool (PU2)	
	The Emotiv Insight could benefit me (PU3)	

items with their corresponding sources. All items were measured using a seven-point Likert-type scale ranging from "strongly agree" to "strongly disagree".

5   Results

Since our data was not distributed joint multivariate normal (cf. Hair et al. 2011), we used the Partial-Least-Squares approach via SmartPLS 3.2.0 (Ringle et al. 2015). With 107 datasets, we met the suggested minimum sample size threshold of "ten times the largest number of structural paths directed at a particular latent construct in the structural model" (Hair et al. 2011, p. 144). To test for significance, we used the integrated Bootstrap routine with 5000 samples (Hair et al. 2011).

5.1   Measurement Model

Tables 2 and 3 present the correlations between constructs along with the Average Variance Extracted (AVE) and Composite Reliability (CR), and our reflective items' factor loadings, respectively: All items loaded high (0.842 or more) and significant (p < 0.001) on their parent factor and, hence, met the suggested threshold of indicator reliability of 0.70 (Hair et al. 2011); AVE and CR were higher than 0.76 and 0.90, respectively, meeting the suggested construct reliability thresholds of 0.50/0.70 (Hair et al. 2009). The loadings from our reflective indicators were

Table 2 Correlations between constructs [AVE (CR) on the diagonal]

	BI	PPR	PU
Behavioral Intention to Use (BI)	0.891 (0.961)	–	–
Perceived Privacy Risk (PPR)	−0.001	0.884 (0.958)	–
Perceived Usefulness (PU)	0.595	−0.121	0.763 (0.906)

Table 3 Reflective items' loadings (T-values)

	BIU	PPR	PU
BI1	0.927 (53.247)	0.033	0.528
BI2	0.946 (86.231)	−0.005	0.586
BI3	0.958 (104.619)	−0.030	0.568
PPR1	0.001	0.877 (8.537)	−0.064
PPR2	0.025	0.973 (12.212)	−0.124
PPR3	−0.028	0.968 (11.718)	−0.131
PU1	0.567	−0.118	0.905 (46.124)
PU2	0.467	−0.018	0.873 (26.618)
PU3	−0.515	−0.167	0.842 (18.058)

Fig. 2 Findings

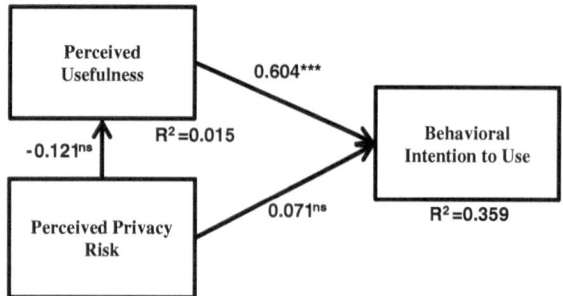

highest for each parent factor and the square root of the AVE of each construct was larger than the absolute value of the construct's correlations with its counterparts, thus indicating discriminant validity (Fornell and Larcker 1981; Hair et al. 2011).

5.2 Structural Model

Figure 2 presents the path coefficients of the previously hypothesized relationships as well as the R^2s of both endogenous variables (*** = p < 0.001; ns = non-significant). Perceived Usefulness was found to have a positive influence on Behavioral Intention to Use ($\beta = 0.604$, p < 0.001), confirming hypothesis 1. However, hypotheses 2 and 3 could not be confirmed since Perceived Privacy Risk had no significant influence on either Behavioral Intention to Use ($\beta = 0.071$, t = 1.094; cf. Ernst 2014; Von Stetten et al. 2011) or Perceived Usefulness ($\beta = -0.121$, t = 1.396).

In summary, our findings indicate that neuroheadsets are at least partly utilitarian technologies whose usage is influenced by Perceived Usefulness. Moreover, Privacy Risks seem to play no part in people's usage of these devices.

6 Conclusions

In this article, we evaluated the potential influence of Perceived Privacy Risk on neuroheadset usage. After collecting 107 complete online questionnaires concerning one specific neuroheadset, Emotiv Insight, and applying a structural equation modeling approach, our findings indicate that neuroheadsets are at least partly utilitarian technologies whose usage is influenced by Perceived Usefulness. However, we were not able to confirm a significant influence of Perceived Privacy Risk on either the Behavioral Intention to Use the headset or its Perceived Usefulness.

Our findings have important practical implications. Indeed, they suggest that neuroheadset manufacturers currently do not need to address people's potential negative perceptions with regards to their privacy and can focus on other matters, such as emphasizing the instrumental benefits of their devices.

Our study has some limitations. First, our empirical findings are only based on one specific neuroheadset: Emotiv Insight. Hence, the results found for this particular neuroheadset might be different in studies that use other headsets. Additionally, Emotiv Insight was not yet available to the general public at the time of this study (June 2015). Hence, our respondents did not have any hands-on experience with the device and could only state their guesses based on our product description as well as on information they might have gathered on their own. Moreover, since we only surveyed English-speaking people, our results might not hold true for non-English speakers. Also, our sample individuals were relatively young (mean: 26.10 years; standard deviation: 6.94). Hence, differences might be found for other age groups. Finally, our survey was only conducted online and, hence, excluded people that do not use the Internet (which might also explain the lack of older people in our sample).

As a next step, we plan to expand our research and address its limitations. More specifically, we first want to identify additional potential influence factors of neuroheadset usage. Following this, and once Emotiv Insight has launched, we plan to equip people in Germany with the device for one week with the instruction of using the device for at least one hour per day. Before and after this hands-on phase, we plan to let our study participants fill out a German-language questionnaire in order to empirically evaluate the influence of each potential influence factor as well as to analyze the potential changes in perception brought on by real-world usage of the device.

References

Ajzen, I. (1991). The theory of planned behavior. *Organizational Behavior and Human Decision Processes, 50*(2), 179–211.

Alarcón-del-Amo, M.-C., Lorenzo-Romero, C., & Gomez-Borja, M.-A. (2012). Analysis of acceptance of social networking sites. *African Journal of Business Management, 6*(29), 8609–8619.

Brief, A. P., & Aldag, R. J. (1977). The intrinsic-extrinsic dichotomy: Toward conceptual clarity. *Academy of Management Review, 2*(3), 496–500.

Chen, R. (2013). Member use of social networking sites—an empirical examination. *Decision Support Systems, 54*(3), 1219–1227.

Davis, F. D. (1989). Perceived usefulness, perceived ease of use, and user acceptance of information technology. *MIS Quarterly, 13*(3), 319–340.

Davis, F. D., Bagozzi, R. P., & Warshaw, P. R. (1989). User acceptance of computer technology: A comparison of two theoretical models. *Management Science, 35*(8), 982–1003.

Dinev, T., & Hart, P. (2006). An extended privacy calculus model for e-commerce transactions. *Information Systems Research, 17*(1), 61–80.

Dowling, G. R., & Staelin, R. (1994). A model of perceived risk and intended risk-handling activity. *Journal of Consumer Research, 21*(1), 119–134.

Egea, J. M. O., & González, M. V. R. (2010). Explaining physicians' acceptance of EHCR systems: An extension of TAM with trust and risk factors. *Computers in Human Behavior, 27*(1), 319–332.

Emotiv. (2015). Insight. http://emotiv.com/insight.php. Accessed July 24, 2015.

Ernst, C.-P. H., Pfeiffer, J., & Rothlauf, F. (2013). Hedonic and utilitarian motivations of social network site adoption. Johannes Gutenberg University Mainz, Working Paper.

Ernst, C.-P. H. (2014). Risk hurts fun: The influence of perceived privacy risk on social network site usage. In *AMCIS 2014 Proceedings*.

Featherman, M. (2001). Extending the technology acceptance model by inclusion of perceived risk. In *AMCIS 2001 Proceedings*. Paper 148.

Featherman, M. S., & Pavlou, P. A. (2003). Predicting e-services adoption: A perceived risk facets perspective. *International Journal of Human-Computer Studies, 59*(4), 451–474.

Fishbein, M., & Ajzen, I. (1975). *Belief, attitude, intention, and behavior: An introduction to theory and research*. Reading, MA: Addison-Wesley.

Fornell, C., & Larcker, D. F. (1981). Evaluating structural equation models with unobservable variables and measurement error. *Journal of Marketing Research, 18*(1), 39–50.

Hair, J. F., Black, W. C., Babin, B. J., & Anderson, R. E. (2009). *Multivariate data analysis* (7th ed.). Upper Saddle River, NJ: Prentice Hall.

Hair, J. F., Ringle, C. M., & Sarstedt, M. (2011). PLS-SEM: Indeed a silver bullet. *Journal of Marketing Theory and Practice, 19*(2), 139–151.

Hu, T., Poston, R. S., & Kettinger, W. J. (2011). Nonadopters of online social network services: Is it easy to have fun yet? *Communications of the Association for Information Systems, 29*(1), 441–458.

Jarvenpaa, S. L., Tractinsky, N., & Vitale, M. (2000). Consumer trust in an Internet store. *Information Technology and Management, 1*(1), 45–71.

Kim, D. J., Ferrin, D. L., & Rao, H. R. (2008). A trust-based consumer decision-making model in electronic commerce: The role of trust, perceived risk, and their antecedents. *Decision Support Systems, 44*(2), 544–564.

Krasnova, H., Spiekermann, S., Koroleva, K., & Hildebrand, T. (2010). Online social networks: Why we disclose. *Journal of Information Technology, 25*(2), 109–125.

Lal, T. N., Schröder, M., Hill, N. J., Preissl, H., Hinterberger, T., Mellinger, J., Bogdan, M., Rosenstiel, W., Hofmann, T., Birbaumer, N., & Schölkopf, B. (2005). A brain computer interface with online feedback based on magnetoencephalography. In *International Conference on Machine Learning 2005 Proceedings*.

Miller, M. (2015). *The Internet of things: How smart TVs, smart cars, smart homes, and smart cities are changing the world*. Indianapolis, IN: Que.

Pavlou, P. A. (2001). Integrating trust in electronic commerce with the technology acceptance model: Model development and validation. In *AMCIS 2001 Proceedings*.

Pavlou, P. A. (2003). Consumer acceptance of electronic commerce: Integrating trust and risk with the technology acceptance model. *International Journal of Electronic Commerce, 7*(3), 101–134.

Peter, J. P., & Ryan, M. J. (1976). An investigation of perceived risk at the brand level. *Journal of Marketing Research, 13*(2), 184–188.

Poslad, S. (2009). *Ubiquitous computing: Smart devices environments and interactions*. Chichester, UK: Wiley.

PwC. (2014). Health wearables: Early days. White Paper.

Ringle, C. M., Wende, S., & Becker, J. -M. (2015). SmartPLS 3. http://www.smartpls.com.

Rose, J., & Fogarty, G. J. (2006). Determinants of perceived usefulness and perceived ease of use in the technology acceptance model: senior consumers' adoption of self-service banking technologies. In *Academy of World Business, Marketing and Management Development Conference Proceedings*.

Sitkin, S. B., & Pablo, A. L. (1992). Reconceptualizing the determinants of risk behavior. *Academy of Management Review, 17*(1), 9–38.

Statista. (2014). Prognose zum Umsatz mit Wearable Technology in Europa von 2013 bis 2018 (in Milliarden Euro). http://de.statista.com/statistik/daten/studie/322222/umfrage/prognose-zum-umsatz-mit-wearable-computing-geraeten-in-europa. Accessed June 29, 2015.

Tan, S. J. (1999). Strategies for reducing consumers' risk aversion in Internet shopping. *Journal of Consumer Marketing, 16*(2), 163–180.

Tehrani, K., & Andrew, M. (2014). Wearable technology and wearable devices: Everything you need to know. http://www.wearabledevices.com/what-is-a-wearable-device. Accessed June 29, 2015.

Vala, N., & Trivedi, K. (2014). Brain computer interface: Data acquisition using non-invasive Emotiv Epoc neuroheadset. *International Journal of Software & Hardware Research in Engineering, 2*(5), 127–130.

Van Der Heijden, H. (2004). User acceptance of hedonic information systems. *MIS Quarterly, 28*(4), 695–704.

Venkatesh, V., Morris, M. G., Davis, G. B., & Davis, F. D. (2003). User acceptance of information technology: Toward a unified view. *MIS Quarterly, 27*(3), 425–478.

Von Stetten, A., Wild, U., & Chrennikow, W. (2011). Adopting social network sites—the role of individual IT culture and privacy concerns. In *AMCIS 2011 Proceedings*. Paper 290.

Westin, A. F. (1968). *Privacy and freedom*. New York, NY: Atheneum.

Wu, Y. A., Ryan, S., & Windsor J. (2009). Influence of social context and affect on individuals' implementation of information security safeguards. In *ICIS 2009 Proceedings*. Paper 70.

Success Comes to Those Who Are Successful: The Influence of Past Product Expectation Confirmation on Smartwatch Usage

Alexander W. Ernst and Claus-Peter H. Ernst

Abstract Due to smartwatches' usual strong functional dependence on other devices from the same manufacturer, we believe that Past Product Expectation Confirmation—which we describe as the extent to which a person believes that his/her expectations were satisfied by a specific manufacturer's product portfolio in the past—influence people's usage of smartwatches. After collecting 229 completed online questionnaires about the Apple Watch, and applying a structural equation modeling approach, our findings indicate that smartwatch usage is positively influenced by Perceived Usefulness. Past Product Expectation Confirmation was found to have a direct positive influence on the Behavioral Intention to Use smartwatches as well as an indirect positive influence on the Behavioral Intention to use smartwatches through Perceived Usefulness. These findings emphasize the importance of having strong product portfolios so that manufacturers can launch equally successful products in the future.

1 Introduction

Wearable devices—i.e., "electronic technologies or computers that are incorporated into items of clothing and accessories which can comfortably be worn on the body" (Tehrani and Andrew 2014)—have gained momentum in the marketplace over the past years. According to IDC (2015), 26.4 million wearable devices were shipped in 2014 and IDC predicts that by 2019 this number will have grown to 155.7 million per year.

One kind of wearable device that has gained particular momentum is the smartwatch: it is projected that 26.1 million devices will be shipped in 2015

A.W. Ernst (✉)
Justus-Liebig-Universität Gießen, Gießen, Germany
e-mail: alexander.ernst@lehramt.uni-giessen.de

C.-P.H. Ernst (✉)
Frankfurt University of Applied Sciences, Frankfurt am Main, Germany
e-mail: cernst@fb3.fra-uas.de

© Springer International Publishing Switzerland 2016   49
C.-P.H. Ernst (ed.), *The Drivers of Wearable Device Usage*,
Progress in IS, DOI 10.1007/978-3-319-30376-5_5

(Statista 2015). Smartwatches are usually worn on the wrist and provide users with multiple utilitarian benefits. However, smartwatches are usually strongly dependent on other devices (e.g., smartphones) from the same manufacturer, and need to be connected via Bluetooth or Wi-Fi to these devices in order to perform most of their functionality (e.g., Apple 2015).

Due to this strong dependence of smartwatches on other devices from the same manufacturer, we believe that people's Past Product Expectation Confirmation— which we describe as the extent to which a person believes that his/her expectations were satisfied by a specific manufacturer's product portfolio in the past (cf. Bhattacherjee 2001)—will influence their purchase and usage of smartwatches. More specifically, we draw from the Expectation Confirmation Theory (e.g., Oliver 1977, 1980) to postulate that if a person's expectations regarding a manufacturer's other products have been satisfied in the past, this will (1) positively influence his/her intention to use a smartwatch made by that particular manufacturer, and (2), positively influence his/her expectations of that manufacturer's smartwatches, i.e., the Perceived Usefulness of their smartwatches [which is a commonly accepted antecedent of technology usage (e.g., Davis et al. 1989; Van der Heijden 2004)].

After collecting 229 complete online questionnaires about one specific smart-watch, the Apple Watch, and applying a structural equation modeling approach, our findings indicate that smartwatch usage is influenced by utilitarian motivations, that is, by their Perceived Usefulness. Past Product Expectation Confirmation was found to have a direct positive influence on the Behavioral Intention to Use smartwatches as well as an indirect influence on Behavioral Intention to Use through Perceived Usefulness. These findings emphasize the importance of having strong product portfolios in order for manufacturers to launch equally successful products in the future.

In the next section, we will present background information on smartwatches, introduce Perceived Usefulness as an influence factor of technologies that provide utilitarian benefits, and also present the theoretical foundations of Past Product Expectation Confirmation. Following this, we will present our research model and research design. We will then reveal and discuss our results before summarizing our findings, presenting their theoretical and practical implications, and providing an outlook on further research.

2 Theoretical Background

2.1 The Role of Perceived Usefulness on Smartwatch Usage

Smartwatches are wearable devices that are typically worn on the wrist. They provide users with multiple utilitarian benefits such as showing the time, cus-tomization of the watches' face, and notifying as well as displaying incoming messages and emails.

Although multiple studies have studied different aspects of wearable devices (e.g., Ariyatum et al. 2005; Bodine and Gemperle 2003; Dvorak 2008; Starner 2001), the factors that drive peoples' smartwatch usage are largely unknown. Indeed, to the best of our knowledge, there is only one article that has studied the factors driving smartwatch usage: Kim and Shin's (2015) findings suggest that Perceived Ease of Use, Affective Quality, Relative Advantage, Mobility, Availability, Subcultural Appeal, Cost, and Perceived Usefulness are influence factors of smartwatch usage.

Generally, utilitarian technologies "aim to provide instrumental value to the user" (Van der Heijden 2004, p. 696). Perceived Usefulness—i.e., "the degree to which a person believes that using a particular system would enhance his or her job [and task] performance" (Davis 1989, p. 320)—centers on the motivations and benefits that are external to the system-user interaction itself, referred to as extrinsic motivations (Brief and Aldag 1977; Van der Heijden 2004). For example, the external benefits/extrinsic motivations of a text-processing program can be to foster a good writing performance in terms of a well-structured and orthographically error-free text (Davis et al. 1989).

Various studies in multiple contexts (e.g., Davis 1989; Kim and Shin 2015) have consistently confirmed that Perceived Usefulness is a central antecedent of utilitarian technologies' usage. In other words, a person can be expected to use smartwatches if he/she believes that they fulfill his expectations with regards to their instrumental benefits, that is, their Perceived Usefulness.

2.2 Past Product Expectation Confirmation

Expectation Confirmation Theory postulates that people's positive or negative disconfirmation of beliefs after performing a certain behavior indirectly influences their intention to re-enact that behavior in the future (Oliver 1977, 1980). For example, people regularly have certain individual expectations regarding a product before the actual purchase. The positive or negative disconfirmation of these perceived expectations after the actual purchase will positively or negatively influence people's future purchase intentions. Drawing from this theory, we believe that smartwatch usage will be influenced by people's Past Product Expectation Confirmation, which we describe as the extent to which a person believes that his/her expectations were satisfied by a specific manufacturer's product portfolio in the past (cf. Bhattacherjee 2001).

More specifically, in order to perform most of their functionality, smartwatches usually need to be connected to other devices (e.g., smartphones) from the same manufacturer via Bluetooth or Wi-Fi (e.g., Apple 2015). As a result, it makes little sense to own just a smartwatch, since its stand-alone capabilities are rather limited. Due to this strong dependence of smartwatches on other devices from the same manufacturer, we believe that people's prior experiences with a manufacturer's other products will influence their purchase and usage of smartwatches. In other words, we will use the Expectation Confirmation Theory not to postulate traditional hypotheses regarding product repurchases (such as 'people will buy an iPhone 7 this year because their iPhone 6s satisfied their expectations last year'), but rather, to postulate hypotheses regarding the usage of a complimentary product (e.g., 'people will buy an Apple Watch today because their iPhones, iPads, etc. satisfied their expectations in the past').

3 Research Model

In the following section, we will present our research model in Fig. 1 and then outline our corresponding hypotheses.

As described earlier, smartwatches provide multiple instrumental benefits to its users such as showing the time and notifying as well as displaying incoming messages and emails (e.g., Apple 2015). Therefore, smartwatches are at least partly utilitarian technologies (cf. Ernst et al. 2013) that provide users with benefits that are external to the system-user interaction itself. Perceived Usefulness is commonly accepted to be an important antecedent of utilitarian technologies' usage (e.g., Davis et al. 1989). We hypothesize that:

H1 *There is a positive influence of Perceived Usefulness on the Behavioral Intention to Use smartwatches.*

The Expectation Confirmation Theory postulates that people's perceived confirmation of their pre-purchase expectations are a positive antecedent of their repurchase intentions (Oliver 1977, 1980). Smartwatches usually need to be connected to other devices from the same manufacturer in order to perform most of their functionalities. Due to this strong dependence on the manufacturer's other devices and drawing from the Expectation Confirmation Theory, we believe that if a person's expectations regarding a manufacturer's other products were satisfied in

Fig. 1 Research model

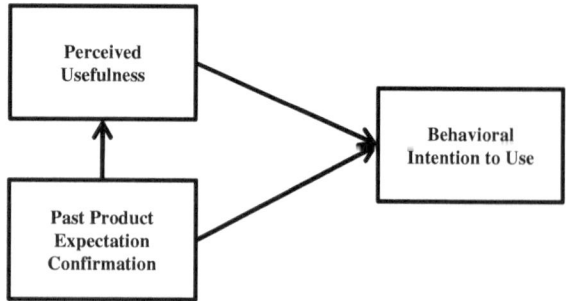

the past, this will positively influence his/her intention to use a smartwatch from that particular manufacturer. We hypothesize that:

H2   *There is a positive influence of Past Product Expectation Confirmation on the Behavioral Intention to Use smartwatches.*

Similarly, we believe that such past expectation satisfaction will positively influence a person's pre-purchase evaluation of smartwatches' functionalities, i.e., their expectation regarding smartwatches' utilitarian benefits. More specifically, the expectations in the Expectation Confirmation Theory "refer to consumers' beliefs about the potential utility that can be derived from a ... [product], which is akin to the notion of perceived usefulness" Bhattacherjee (2001, p. 204). Previous confirmed or unconfirmed expectations regarding a product are able to change people's future expectations (Bhattacherjee 2001; cf. Oliver 1980). Indeed, expectations "may be adjusted higher ... if customers realize that their initial expectations were unrealistically low. Likewise, unreasonably high initial expectations ... may be lowered ... as some of those expectations are disconfirmed" (Bhattacherjee 2001, p. 204). In other words and adjusted to our context, if people's expectations were confirmed by a manufacturer's product portfolio in the past, their expectation regarding the usefulness of the manufacturer's smartwatches will be positively affected. We hypothesize that:

H3   *There is a positive influence of Past Product Expectation Confirmation on the Perceived Usefulness of smartwatches.*

4   Research Design

4.1   Data Collection

To empirically evaluate our research model, we collected 229 completed German-language online questionnaires about one specific smartwatch: the Apple Watch. At the beginning of the questionnaire, we gave a short description of the Apple Watch, including official images and an explanation of its general functionalities: The Apple Watch was released in selected countries including Germany and the US on April 24th, 2015 (Apple 2015). In order to perform most of its functionality, the Apple Watch needs to be connected to an iPhone via Bluetooth or Wi-Fi. Among its most prominently advertised functions are the collection, monitoring and storing of physical activity data and health-related data such as heart rate, miles walked, time of activity, and calories burned as well as the notification and display of incoming messages and emails, the customization of the watches' face, the sending and receiving of small doodles, and the possibility of sharing one's own heartbeat.

152 of our respondents were female (66.38 %) and 77 were male (33.62 %). The average age was 28.00 years (standard deviation: 9.29). 1 respondent was unemployed (0.4 %), 3 respondents were apprentices (1.3 %), 2 were pupils (0.9 %), 151 were students (65.9 %), 65 were currently employed (28.4 %), and 7 selected "other" as a description of themselves (3.1 %).

4.2 Measurement

We adapted existing reflective scales to our context in order to measure the Behavioral Intention to Use and Perceived Usefulness. For Past Product Expectation Confirmation, we developed three of our own reflective items. In order to do this, we based ourselves on the existing Expectation Confirmation scale from Bhattacherjee (2001) and consulted several researchers from our department throughout the development process. Table 1 presents the resulting reflective items with their corresponding sources. All items were measured using a seven-point Likert-type scale ranging from "strongly agree" to "strongly disagree".

Table 1 Items of our measurement model

Construct	Items	Adapted from
Behavioral Intention to Use	I intend to use an Apple Watch in the next 6 months (BI1)	Hu et al. (2011) Venkatesh et al. (2003)
	I expect I will use an Apple Watch in the near future (BI2)	
	In the future, I am very likely to use an Apple Watch (BI3)	
Past Product Expectation Confirmation	Usually, Apple satisfies my expectations (PPEC1)	Created by ourselves cf. Bhattacherjee (2001)
	Apple has regularly met my expectations in the past (PPEC2)	
	Until now, Apple always lived up to my expectations (PPEC3)	
Perceived Usefulness	Overall, an Apple Watch is useful (PU1)	Alarcón-del-Amo et al (2012) cf. Ernst et al. (2013)
	I consider that the Apple Watch is useful to me (PU2)	
	The Apple Watch benefits me (PU3)	

5 Results

Since our data was not distributed joint multivariate normal (cf. Hair et al. 2011), we used the Partial-Least-Squares approach via SmartPLS 3.2.0 (Ringle et al. 2015). With 229 datasets, we met the suggested minimum sample size threshold of "ten times the largest number of structural paths directed at a particular latent construct in the structural model" (Hair et al. 2011, p. 144). To test for significance, we used the integrated Bootstrap routine with 5,000 samples (Hair et al. 2011).

In the following section, we will evaluate our measurement model. Indeed, we will examine the indicator reliability, the construct reliability, and the discriminant validity of our reflective constructs. Finally, we will present the results of our structural model.

5.1 Measurement Model

Tables 2 and 3 present the correlations between constructs along with the Average Variance Extracted (AVE) and Composite Reliability (CR), and our reflective items' factor loadings, respectively: All items loaded high (0.938 or more) and significant ($p < 0.001$) on their parent factor and, hence, met the suggested threshold of indicator reliability of 0.70 (Hair et al. 2011); AVE and CR were higher than

Table 2 Correlations between constructs [AVE (CR) on the diagonal]

	BI	PPEC	PU
Behavioral Intention to Use (BI)	0.927 (0.974)		
Past Product Expectation Confirmation (PPEC)	0.486	0.940 (0.979)	
Perceived Usefulness (PU)	0.555	0.529	0.894 (0.962)

Table 3 Reflective items' loadings (T-values)

	BI	PPEC	PU
BI1	0.959 (74.409)	0.441	0.505
BI2	0.973 (59.727)	0.474	0.530
BI3	0.955 (61.439)	0.485	0.564
PPEC1	0.477	0.973 (118.931)	0.545
PPEC2	0.470	0.964 (108.828)	0.496
PPEC3	0.467	0.972 (162.027)	0.496
PU1	0.518	0.513	0.956 (147.521)
PU2	0.524	0.493	0.942 (98.932)
PU3	0.531	0.493	0.938 (90.976)

Fig. 2 Findings

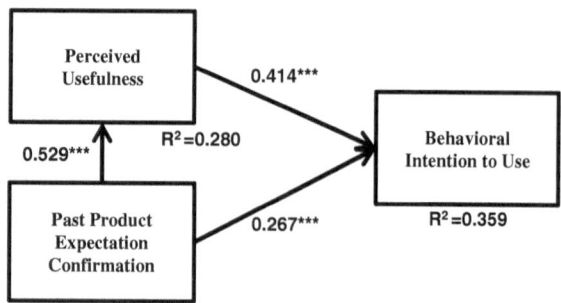

0.89 and 0.96, respectively, meeting the suggested construct reliability thresholds of 0.50/0.70 (Hair et al. 2009). The loadings from our reflective indicators were highest for each parent factor and the square root of the AVE of each construct was larger than the absolute value of the construct's correlations with its counterparts, thus indicating discriminant validity (Fornell and Larcker 1981; Hair et al. 2011).

5.2 Structural Model

Figure 2 presents the path coefficients of the previously hypothesized relationships as well as the R^2s of both endogenous variables (*** = $p < 0.001$).

All hypotheses of our research model were confirmed: Perceived Usefulness ($\beta = 0.414$, $p < 0.001$) and Past Product Expectation Confirmation ($\beta = 0.267$, $p < 0.001$) had a significant positive influence on Behavioral Intention to Use. Additionally, Past Product Expectation Confirmation had a significant positive influence on Perceived Usefulness ($\beta = 0.529$, $p < 0.001$).

Overall, our research model included two predecessors of Behavioral Intention to Use (Perceived Usefulness and Past Product Expectation Confirmation) and one predecessor of Perceived Usefulness (Past Product Expectation Confirmation). By taking this into account, the explanatory power of our structural model is good, since it explains 35.9 % of the variances of Behavioral Intention to Use and 28.0 % of the variances of Perceived Usefulness.

6 Conclusions

In this article, we evaluated the influence of Past Product Expectation Confirmation on smartwatch usage. After collecting 229 completed online questionnaires and applying a structural equation modeling approach, our findings suggest that smartwatches are at least partly utilitarian technologies, whose usage is positively influenced by Perceived Usefulness. Past Product Expectation Confirmation was found to positively influence the Behavioral Intention to Use smartwatches directly, and indirectly through Perceived Usefulness.

Our findings have important practical implications. Indeed, they suggest that smartwatch manufacturers need to not only emphasize the utilitarian benefits of their smartwatches, but also emphasize the utilitarian benefits of their other devices in order to ultimately achieve greater smartwatch market penetration. Moreover, our results indicate that smartwatch success may be better achieved by manufacturers who already have successful product portfolios than by manufacturers who do not. In other words, our findings emphasize the importance of strong product portfolios in order for manufacturers to launch equally successful products in the future, suggesting that success comes rather to those who are already successful than to those that are not.

Our study has some limitations. First, our empirical findings are based on only one specific smartwatch: the Apple Watch. Therefore, there might be differences between this particular smartwatch and other smartwatches, especially in the case of smartwatches that do not depend on devices from the same manufacturer but rather can be used with devices from multiple manufacturers. Moreover, since we only surveyed German-speaking people, our results might not hold true for non-German speaking people. In addition, our sample individuals were relatively young (mean: 28.00 years; standard deviation: 9.29). Hence, differences might be found for other age groups. Finally, our survey was only conducted online and, hence, excluded people that do not use the Internet (which might also explain the lack of older people in our sample).

In the case of this article, we studied the influence of Past Product Expectation Confirmation on the usage of a manufacturer's product, in the case where the new product (a smartwatch) depends on the older product (a smartphone, tablet, etc.) to function. As a next step, we plan to further evaluate the influence of Past Product Expectation Confirmation on product usage. More specifically, Past Product Expectation Confirmation might influence the usage of a manufacturer's products even though both the new product and the products previously used by the customer are stand-alone technologies, i.e., neither product is dependent on any others in order to function. For example, people's intention to buy a fridge might also be driven by their past product expectation confirmation with regards to the specific manufacturer's smartphones, hair dryers, etc. Hence, we plan to examine the implications of Past Product Expectation Confirmation in a variety of contexts in subsequent studies.

References

Alarcón-del-Amo, M.-C., Lorenzo-Romero, C., & Gomez-Borja, M.-A. (2012). Analysis of acceptance of social networking sites. *African Journal of Business Management, 6*(29), 8609–8619.

Apple. (2015). Watch. http://www.apple.com/watch. Accessed September 9, 2015.

Ariyatum, B., Holland, R., Harrison, D., & Kazi, T. (2005). The future design direction of smart clothing development. *Journal of the Textile Institute, 96*(4), 199–210.

Bhattacherjee, A. (2001). An empirical analysis of the antecedents of electronic commerce service continuance. *Decision Support Systems, 32*(2), 201–214.

Bodine, K., & Gemperle, F. (2003). Effects of functionality of perceived comfort of wearables. *IEEE International Symposium on Wearable Computers 2003 Proceedings.*

Brief, A. P., & Aldag, R. J. (1977). The intrinsic-extrinsic dichotomy: Toward conceptual clarity. *Academy of Management Review, 2*(3), 496–500.

Davis, F. D. (1989). Perceived usefulness, perceived ease of use, and user acceptance of information technology. *MIS Quarterly, 13*(3), 319–340.

Davis, F. D., Bagozzi, R. P., & Warshaw, P. R. (1989). User acceptance of computer technology: A comparison of two theoretical models. *Management Science, 35*(8), 982–1003.

Dvorak, J. L. (2008). *Moving Wearables into the Mainstream: Taming the Borg.* New York, NY: Springer.

Ernst, C.-P.H., Pfeiffer, J., & Rothlauf, F. (2013). *Hedonic and utilitarian motivations of social network site adoption.* Working Paper. Johannes Gutenberg University Mainz.

Fornell, C., & Larcker, D. F. (1981). Evaluating structural equation models with unobservable variables and measurement error. *Journal of Marketing Research, 18*(1), 39–50.

Hair, J. F., Black, W. C., Babin, B. J., & Anderson, R. E. (2009). *Multivariate data analysis* (7th ed.). Upper Saddle River, NJ: Prentice Hall.

Hair, J. F., Ringle, C. M., & Sarstedt, M. (2011). PLS-SEM: Indeed a silver bullet. *Journal of Marketing Theory and Practice, 19*(2), 139–151.

Hu, T., Poston, R. S., & Kettinger, W. J. (2011). Nonadopters of online social network services: Is it easy to have fun yet? *Communications of the Association for Information Systems, 29*(1), 441–458.

IDC. (2015). Worldwide wearables market forecast to grow 173.3 % in 2015 with 72.1 million units to be shipped, according to IDC. http://www.idc.com/getdoc.jsp?containerId=prUS25696715. Accessed May 5, 2015.

Kim, K. J., & Shin, D.-H. (2015). An acceptance model for smart watches: Implications for the adoption of future wearable technology. *Internet Research, 25*(4), 527–541.

Oliver, R. L. (1977). Effect of expectation and disconfirmation on postexposure product evaluations: An alternative interpretation. *Journal of Applied Psychology, 62*(4), 480–486.

Oliver, R. L. (1980). A cognitive model of the antecedents and consequences of satisfaction decisions. *Journal of Marketing Research, 17*(4), 460–469.

Ringle, C.M., Wende, S., & Becker, J.-M. (2015). SmartPLS 3. http://www.smartpls.com.

Starner, T. (2001). The challenges of wearable computing: Part 2. *IEEE Micro, 21*(4), 54–67.

Statista. (2015). Forecast unit sales of smartwatches worldwide by region from 2014 to 2015. http://www.statista.com/statistics/413248/smartwatch-worldwide-unit-sales-region. Accessed September 9, 2015.

Tehrani, K., & Andrew, M. (2014). Wearable technology and wearable devices: Everything you need to know. http://www.wearabledevices.com/what-is-a-wearable-device. Accessed June 6, 2015.

Van der Heijden, H. (2004). User acceptance of hedonic information systems. *MIS Quarterly, 28*(4), 695–704.

Venkatesh, V., Morris, M. G., Davis, G. B., & Davis, F. D. (2003). User acceptance of information technology: Toward a unified view. *MIS Quarterly, 27*(3), 425–478.

How Design Influences Headphone Usage

Patrick Reinelt, Shewit Hadish and Claus-Peter H. Ernst

Abstract Headphones are some of the most popular wearable devices. However, the factors driving their usage are largely unknown. In this article, we postulate a positive influence of Perceived Design Aesthetics on headphone usage. After collecting 125 completed online questionnaires about one specific pair of headphones, Beats by Dr. Dre Studio Wireless, and applying a structural equation modeling approach, our findings indicate that headphones are at least partly hedonic technologies whose usage is influenced by Perceived Enjoyment. Furthermore, although we could not confirm a direct positive influence of Perceived Design Aesthetics on the Actual System Use of headphones, we confirmed an indirect influence of Perceived Design Aesthetics on Actual System Use through Perceived Enjoyment. These findings suggest that headphone manufacturers need to emphasize the hedonic character of their devices, and that designing their devices should be undertaken with the utmost care.

1 Introduction

Wearable devices—i.e., "electronic technologies or computers that are incorporated into items of clothing and accessories which can comfortably be worn on the body" (Tehrani and Andrew 2014)—have gained momentum in the marketplace over the past years (e.g., Lopez et al. 2010). One of the most popular forms of wearable devices are headphones. It has been predicted that 290.9 million units will be sold in 2015 alone (Statista 2015). However, the factors driving the usage of headphones are largely unknown.

Since headphones are worn on, over, or inside the ear, they are more or less visible to others (Reeves and Nass 1996). Hence, similar to items of clothing, the

P. Reinelt · S. Hadish · C.-P.H. Ernst (✉)
Frankfurt University of Applied Sciences, Frankfurt am Main, Germany
e-mail: cernst@fb3.fra-uas.de

© Springer International Publishing Switzerland 2016
C.-P.H. Ernst (ed.), *The Drivers of Wearable Device Usage*,
Progress in IS, DOI 10.1007/978-3-319-30376-5_6

aesthetical characteristics of the headphones might play an important role when customers are deciding whether to use a specific pair of headphones or not (cf. Cooper and Kleinschmidt 1987). We thus postulate a positive influence of Perceived Design Aesthetics on headphone usage.

After collecting 125 completed online questionnaires about one specific pair of headphones, Beats by Dr. Dre Studio Wireless, and applying a structural equation modeling approach, our findings indicate that headphones are at least partly hedonic technologies whose usage is influenced by Perceived Enjoyment. Furthermore, although we could not confirm a direct positive influence of Perceived Design Aesthetics on the Actual System Use of headphones, we confirmed an indirect positive influence of Perceived Design Aesthetics on Actual System Use through Perceived Enjoyment. These findings suggest that headphone manufacturers need to emphasize the hedonic character of their devices as well as put great care into the design of these devices.

In the next section, we will present background information on headphones, introduce Perceived Enjoyment as an influence factor of hedonic technologies, and also present the theoretical foundations of Perceived Design Aesthetics. Following this, we will present our research model and research design. We will then reveal and discuss our results before summarizing our findings, presenting their theoretical as well as practical implications, and providing an outlook on further research.

2 Theoretical Background

2.1 The Role of Perceived Enjoyment on Headphone Usage

Headphones are loudspeakers that are worn on, over, or inside the ear. They enable people to hear sound such as music and speech and can be connected to an audio source via a cable, or wirelessly.

Listening to music, watching movies and TV shows, playing video games, etc., are usually considered leisure activities and are, hence, generally accepted to provide people with hedonic benefits such as enjoyment, pleasure, excitement, etc. (cf. Hirschman and Holbrook 1982). It is highly probable that headphone usage is regularly linked to these hedonic contexts, making headphones at least partly hedonic technologies (cf. Ernst et al. 2013).

In general, hedonic technologies "aim to provide self-fulfilling value to the user, … [which] is a function of the degree to which the user experiences fun when using the system" (Van der Heijden 2004, p. 696). Various studies in multiple contexts have consistently confirmed that Perceived Enjoyment—i.e., "the extent to which the activity of using a specific system is perceived to be enjoyable in its own right, aside from any performance consequences resulting from system use" (Venkatesh 2000, p. 351)—is a central antecedent of hedonic technology usage (e.g., Van der Heijden 2004). By applying these findings to our contexts, a person can be expected to use headphones if he/she believes that they fulfill his/her expectations with regards to their hedonic benefits, such as enjoyment and pleasure (e.g., Van der Heijden 2004).

2.2 Design Aesthetics

People like beautiful things. In fact, the physical appearance of a product is generally accepted to influence people's purchase decisions. For example, Cooper and Kleinschmidt (1987) studied the market entry performance of new products and confirmed that physical appearance significantly influences sales success. As a consequence, manufacturers regularly pay special attention to a product's physical appearance in order to gain the upper hand against their competitors (Kotler and Rath 1984; Russell and Pratt 1980; Whitney 1988).

In terms of wearable devices, the physical appearance of the product might be an even more important influence factor when users are deciding whether or not to use it. Indeed, wearable devices are regularly worn in a manner that is visible to other people (e.g., Reeves and Nass 1996; Tractinsky et al. 2000), and the devices' color, shape and size can often change the entire appearance of a person (Sonderegger and Sauer 2010). Hence, people might pay special attention to the aesthetics and design of wearable devices (Bodine and Gemperle 2003). Additionally, in highly competitive markets with more or less homogenous products, the device's appearance can define the product, for example, with regards to its durability, technical sophistication, or prestige, resulting in an increase in customers' attention and awareness of both the product and the brand (Berkowitz 1987; Forty 1986). As a result, the physical appearance of headphones might be an important factor driving their usage. Indeed, multiple headphone manufacturers already pay particular attention to their headphones' design and specifically emphasize the device's aesthetics when marketing their product (e.g., Beats 2015).

In the literature, different terms have been used to describe the physical appearance of a product. While some researchers have used the term 'design' (e.g., Bloch 1995; Cooper and Kleinschmidt 1987), others have used the term 'aesthetic' (e.g., Holbrook 1980; Lavie and Tractinsky 2004). In this article, we describe the extent to which a person likes the design of a product as 'Perceived Design Aesthetics' (cf. Ivanov and Cyr 2014). In the following sections, we will postulate hypotheses regarding the influence of Perceived Design Aesthetics on peoples' headphone usage.

3   Research Model

In the following section, we will present our research model in Fig. 1 and then outline our corresponding hypotheses.

As described earlier, headphones will be regularly be used in fun, pleasurable and exciting contexts. Therefore, headphones can be seen as at least partly hedonic technologies that can provide positive feelings and experiences for their users (cf. Van der Heijden 2004). Perceived Enjoyment has been shown to be an important antecedent of hedonic technology usage (e.g., Ernst et al. 2013; Van der Heijden 2004). We hypothesize that:

H1   *There is a positive influence of Perceived Enjoyment on the Actual System Use of headphones.*

The physical appearance of a product positively influences people's purchase decisions (e.g., Cooper and Kleinschmidt 1987). Headphones are worn in a manner that is visible to other people; moreover, headphones can define a person's entire appearance (Sonderegger and Sauer 2010). Hence, people might choose the headphones that they find most visually appealing (cf. Bodine and Gemperle 2003). Moreover, Yamamoto and Lambert (1994) suggest that people are often not able to make an objective decision in complex purchase situations and that the physical appearance of the product can thus act as the main factor driving the actual choice. We hypothesize that:

H2   *There is a positive influence of Perceived Design Aesthetics on the Actual System Use of headphones.*

Also, aesthetics have been shown to be an important influence factor of peoples' subjective feelings (e.g. Thüring and Mahlke 2007). More specifically, aesthetics cause pleasurable subjective experiences (cf. Bloch 1995; Reber et al. 2004). Perceived Enjoyment reflects positive experiences and feelings (Brief and Aldag

Fig. 1   Research Model

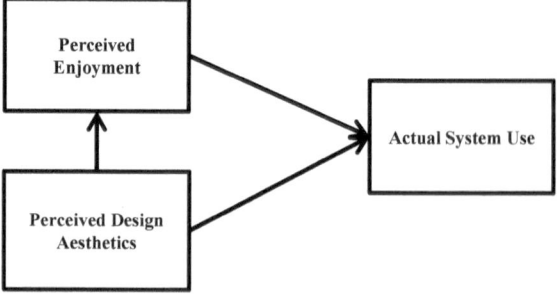

1977; Van der Heijden 2004; Venkatesh et al. 2012). Consequently, we believe that Perceived Design Aesthetics has a positive influence on Perceived Enjoyment in the context of headphone use. We hypothesize that:

H3 *There is a positive influence of Perceived Design Aesthetics on the Perceived Enjoyment of headphones.*

4 Research Design

4.1 Data Collection

To empirically evaluate our research model, we collected 125 completed German-language online questionnaires about one specific pair of headphones, Beats by Dr. Dre Studio Wireless (Beats 2015). These headphones enable users to listen to music wirelessly and to make hands-free calls. Moreover, they provide active external noise cancelling and they have a battery life of up to 12 h. Finally, they are available in various different colors. At the beginning of the questionnaire, we provided a short description of the headphones, including an explanation of their general functionalities as well as two images, one showing the different color options available, and one depicting a person wearing the headphones in his daily routine.

64 of our respondents were male (51.20 %) and 61 were female (48.80 %). The average age was 24.78 years (standard deviation: 5.50). 3 respondents were unemployed (2.4 %), 3 were apprentices (2.4 %), 2 were pupils (1.6 %), 22 were currently employed (17.6 %), 4 were self-employed (3.2), and 91 were students (72.8 %).

4.2 Measurement

We adapted existing reflective scales to our context in order to measure the Actual System Use of the headphones and their Perceived Enjoyment. For Perceived Design Aesthetics, we developed three of our own reflective items and consulted several researchers from our department throughout the development process. Table 1 presents the resulting reflective items with their corresponding sources. Actual System Use was measured in the same manner as in Davis et al. (1989, p. 991), and all other items were measured using a seven-point Likert-type scale ranging from "strongly agree" to "strongly disagree".

Table 1 Items of our measurement model

Construct	Items (Labels)	Source/Adapted from
Actual System Use	On average, how often do you use Beats by Dre[a] (AU1)	Davis et al. (1989)
	How frequently do you use Beats by Dre (AU2)	
Perceived Design Aesthetics	The look of Beats by Dre is appealing to me (PDA1)	created by ourselves
	I like the design of Beats by Dre (PDA2)	
	I think that Beats by Dre have a nice design (PDA3)	
Perceived Enjoyment	I find using Beats by Dre to be enjoyable (PE1)	Davis et al. (1992)
	Using Beats by Dre is pleasant (PE2)	
	I have fun using Beats by Dre (PE3)	

[a]We did not use the full name of the Beats by Dr. Dre Studio Wireless or an abbreviation of the name; instead, we chose to simply refer to the headphones as Beats by Dre in both our introductory text and items

5 Results

Since our data was not distributed joint multivariate normal (cf. Hair et al. 2011), we used the Partial-Least-Squares approach via SmartPLS 3.2.0 (Ringle et al. 2015). With 125 datasets, we met the suggested minimum sample size threshold of "ten times the largest number of structural paths directed at a particular latent construct in the structural model" (Hair et al. 2011, p. 144). To test for significance, we used the integrated Bootstrap routine with 5,000 samples (Hair et al. 2011).

In the following section, we will evaluate our measurement model. Indeed, we will examine the indicator reliability, the construct reliability, and the discriminant validity of our reflective constructs. Finally, we will present the results of our structural model.

5.1 Measurement Model

Tables 2 and 3 present the correlations between constructs along with the Average Variance Extracted (AVE) and Composite Reliability (CR), as well as our reflective items' factor loadings, respectively: All items loaded high (0.957 or more) and significant ($p < 0.001$) on their parent factor and, hence, met the suggested threshold of indicator reliability of 0.70 (Hair et al. 2011). AVE and CR were higher than 0.92 and 0.97, respectively, meeting the suggested construct reliability thresholds of 0.50/0.70 (Hair et al. 2009). The loadings from our reflective indicators were

Table 2 Correlations between constructs [AVE (CR) on the diagonal]

	AU	PDA	PE
Actual System Usage (AU)	0.993 (0.997)		
Perceived Design Aesthetics (PDA)	0.537	0.948 (0.982)	
Perceived Enjoyment (PE)	0.691	0.791	0.928 (0.975)

Table 3 Reflective items' loadings (T-values)

	AU	PDA	PE
AU1	0.997 (544.790)	0.541	0.700
AU2	0.997 (468.974)	0.529	0.678
PDA1	0.556	0.970 (112.258)	0.767
PDA2	0.499	0.971 (150.013)	0.772
PDA3	0.511	0.980 (183.989)	0.769
PE1	0.669	0.726	0.972 (155.043)
PE2	0.681	0.778	0.957 (100.342)
PE3	0.648	0.779	0.960 (115.359)

highest for each parent factor and the square root of the AVE of each construct was larger than the absolute value of the construct's correlations with its counterparts, thus indicating discriminant validity (Fornell and Larcker 1981; Hair et al. 2011).

5.2 Structural Model

Figure 2 presents the path coefficients of the previously hypothesized relationships as well as the R^2s of both endogenous variables (*** = $p < 0.001$; ns = non-significant).

Hypothesis 2 was not confirmed since Perceived Design Aesthetics had no significant influence on Actual System Use ($\beta = -0.026$, t = 0.444). However, Perceived Design Aesthetics was found to have a positive influence on Perceived

Fig. 2 Findings

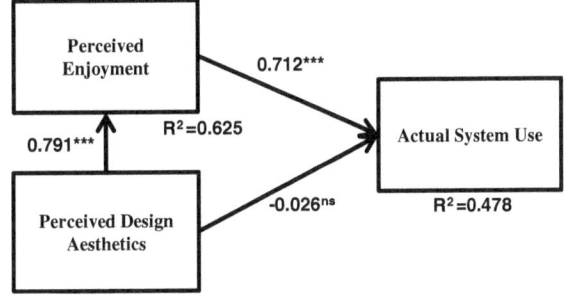

Enjoyment ($\beta = 0.791$, $p < 0.001$), which, in turn, was found to have a positive influence on Actual System Use ($\beta = 0.712$, $p < 0.001$), confirming hypotheses 3 and 1, respectively.

Our research model included two predecessors of Actual System Use (Perceived Enjoyment and Perceived Design Aesthetics), and one predecessor of Perceived Enjoyment (Perceived Design Aesthetics). By taking this into account, the explanatory power of our structural model is good, since it explains 47.8 % of the variances of Actual System Use as well as 62.5 % of the variances of Perceived Enjoyment.

6 Conclusion

In this article, we evaluated the potential influence of Perceived Design Aesthetics on headphone usage. After collecting 125 completed online questionnaires about one specific headphone, Beats by Dr. Dre Studio Wireless, and applying a structural equation modeling approach, our findings indicated that headphones are at least partly hedonic technologies whose usage is influenced by Perceived Enjoyment. Furthermore, although we could not confirm a direct positive influence of Perceived Design Aesthetics on the Actual System Use of headphones, we confirmed an indirect positive influence of Perceived Design Aesthetics on Actual System Use through Perceived Enjoyment.

Our findings have important practical implications. Indeed, they suggest that headphone manufacturers need to emphasize the hedonic character of their devices as well as put great care into the design of their devices. For example, market research carried out during the product development process might help manufacturers understand the preferences of potential customer groups. Also, manufacturers could choose to offer multiple variations with regards to color, shape and size in order to appeal to different customer preferences. Finally, marketing efforts could concentrate on promoting the aesthetics aspects of the headphones.

Our study has some limitations. First, our empirical findings are based only on one specific set of headphones: Beats by Dr. Dre Studio Wireless. Hence, the results found for these particular headphones might be different in studies that use other headphones. Moreover, since we only surveyed German-speaking people, our results might not hold true for non-German speakers. Also, our sample individuals were relatively young (mean: 24.78 years; standard deviation: 5.50). Hence, differences might be found for other age groups.

As a next step, we plan to expand our research and address its limitations. More specifically, we want to rollout our survey to a greater number of countries around the world using different headphones, in order to evaluate for potential differences between countries and devices. We also want to broaden our analysis by taking into account both a hedonic and a utilitarian perspective. More specifically, we plan to include Perceived Sound Quality as well as Perceived Usefulness into our research model in order to evaluate whether Design or Sound Quality is the most important factor driving consumer's headphone usage.

References

Beats. (2015). Studio wireless. http://www.beatsbydre.com/headphones/beats-studio-wireless.html. Accessed July 30, 2015.

Berkowitz, M. (1987). Product shape as a design innovation strategy. *Journal of Product Innovation Management, 4*(4), 274–283.

Bloch, P. H. (1995). Seeking the ideal form: Product design and consumer response. *Journal of Marketing, 59*(3), 16–29.

Bodine, K., & Gemperle, F. (2003). Effects of functionality of perceived comfort of wearables. In *IEEE International Symposium on Wearable Computers 2003 Proceedings.*

Brief, A. P., & Aldag, R. J. (1977). The intrinsic-extrinsic dichotomy: Toward conceptual clarity. *Academy of Management Review, 2*(3), 496–500.

Cooper, R. G., & Kleinschmidt, E. (1987). New products: What separates winners from losers? *Journal of Product Innovation Management, 4*(12), 169–184.

Davis, F. D., Bagozzi, R. P., & Warshaw, P. R. (1989). User acceptance of computer technology: A comparison of two theoretical models. *Management Science, 35*(8), 983–1003.

Davis, F. D., Bagozzi, R. P., & Warshaw, P. R. (1992). Extrinsic and intrinsic motivation to use computers in the workplace. *Journal of Applied Social Psychology, 22*(14), 1111–1132.

Ernst, C.-P.H., Pfeiffer, J., & Rothlauf, F. (2013). *Hedonic and utilitarian motivations of social network site adoption.* Johannes Gutenberg University Mainz, Working paper.

Fornell, C., & Larcker, D. F. (1981). Evaluating structural equation models with unobservable variables and measurement error. *Journal of Marketing Research, 18*(1), 39–50.

Forty, A. (1986). *Objects of desire.* New York: Pantheon Books.

Hair, J. F., Black, W. C., Babin, B. J., & Anderson, R. E. (2009). *Multivariate data analysis* (7th ed.). Upper Saddle River: Prentice Hall.

Hair, J. F., Ringle, C. M., & Sarstedt, M. (2011). PLS-SEM: Indeed a silver bullet. *Journal of Marketing Theory and Practice, 19*(2), 139–151.

Hirschman, E. C., & Holbrook, M. B. (1982). Hedonic consumption: Emerging concepts, methods and propositions. *Journal of Marketing, 46*(3), 92–101.

Holbrook, M. B. (1980). Some preliminary notes on research in consumer aesthetics. *Advances in Consumer Research, 7*(1), 104–108.

Ivanov, A., & Cyr, D. (2014). Satisfaction with outcome and process from web-based meetings for idea generation and selection: The roles of instrumentality, enjoyment, and interface design. *Telematics and Informatics, 31*(4), 543–558.

Kotler, P., & Rath, G. A. (1984). Design: A powerful but neglected strategic tool. *Journal of Business Strategy, 5*(1), 16–21.

Lavie, T., & Tractinsky, N. (2004). Assessing dimensions of perceived visual aesthetics of web sites. *International Journal of Human-Computer Studies, 60*(3), 269–298.

Lopez, G., Shuzo, M., & Yamada, I. (2010). New healthcare society supported by wearable sensors and information mapping based services. *International Journal of Networking and Virtual Organizations, 15*(2), 1–15.

Reber, R., Schwarz, N., & Winkielman, P. (2004). Processing fluency and aesthetic pleasure: Is beauty in the perceiver's processing experience? *Personality and Social Psychology Review, 8*(4), 364–382.

Reeves, B., & Nass, C. (1996). *The media equation: How people treat computers, television, and new media like real people and places.* New York: Cambridge University Press.

Ringle, C.M., Wende, S., & Becker, J.-M. (2015). SmartPLS 3. http://www.smartpls.com.

Russell, J. A., & Pratt, G. (1980). A description of the affective quality attributed to environments. *Journal of Personality and Social Psychology, 38*(2), 311–322.

Sonderegger, A., & Sauer, J. (2010). The influence of design aesthetics in usability testing: Effects on user performance and perceived usability. *Applied Ergonomics, 41*(3), 403–410.

Statista. (2015). *Global unit sales of headphones and headsets from 2010 to 2015* (in millions). http://www.statista.com/statistics/327000/worldwide-sales-headphones-headsets. Accessed July 31, 2015.

Tehrani, K., & Andrew, M. (2014). *Wearable technology and wearable devices: Everything you need to know*. http://www.wearabledevices.com/what-is-a-wearable-device. Accessed June 29, 2015.

Thüring, M., & Mahlke, S. (2007). Usability, aesthetics and emotions in human-technology interaction. *International Journal of Psychology, 42*(4), 253–264.

Tractinsky, N., Ikar, D., & Katz, A. S. (2000). What is beautiful is usable. *Interacting with Computers, 13*(1), 127–145.

Van der Heijden, H. (2004). User acceptance of hedonic information systems. *MIS Quarterly, 28*(4), 695–704.

Venkatesh, V. (2000). Determinants of perceived ease of use: Integrating control, intrinsic motivation, and emotion into the technology acceptance model. *Information Systems Research, 11*(4), 342–365.

Venkatesh, V., Thong, J. Y. L., & Xu, X. (2012). Consumer acceptance and use of information technology: Extending the unified theory of acceptance and use of technology. *MIS Quarterly, 36*(1), 157–178.

Whitney, D. E. (1988). Manufacturing by design. *Harvard Business Review, 66*(4), 83–90.

Yamamoto, M., & Lambert, D. R. (1994). The impact of product aesthetics on the evaluation of industrial products. *Journal of Product Innovation Management, 11*(4), 309–324.